BASICS OF THERMODYNAMICS

BASICS OF THERMODYNAMICS

Alimuddin Khan

ANMOL PUBLICATIONS PVT. LTD.
NEW DELHI - 110 002 (INDIA)

ANMOL PUBLICATIONS PVT. LTD.
H.O.: 4374/4B, Ansari Road, Daryaganj,
New Delhi-110 002 (India)
Ph.: 23278000, 23261597
B.O.: No. 1015, Ist Main Road, BSK IIIrd Stage
IIIrd Phase, IIIrd Block,
Bangalore - 560 085 (India)
Visit us at: www.anmolpublications.com

Basics of Thermodynamics

First Published, 2008

ISBN 978-81-261-3566-0

PRINTED IN INDIA

Printed at Mehra Offset Press, Delhi.

Contents

Preface

The fundamental aspects of thermodynamics are presented in a simple compact way. The equations derived are illustrated by numerous (111) examples, often direct application of the relations just obtained. The (four) laws of thermodynamics are presented and illustrated. The need to define thermodynamic temperature, the meaning of auxiliary thermodynamic functions, the origin, usefulness and use of partial molar quantities are all examined. Gaseous systems, phase equilibria and chemical reactions are quantitatively treated. It is shown how chemical reactions can provide work. Ideal and non ideal solutions are presented with the various standard states and activity coefficients. This book will be of use to a wide audience of students and professionals in the fields of Physics, Engineering, Materials Science

Author

Chapter 1

Heat

FEELING AND SEEING TEMPERATURE CHANGES

Within some reasonable temperature range, we can get a rough idea how warm something is by touching it. But this can be unreliable—if you put one hand in cold water, one in hot, then plunge both of them into lukewarm water, one hand will tell you it's hot, the other will feel cold. For something too hot to touch, we can often get an impression of how hot it is by approaching and sensing the radiant heat. If the temperature increases enough, it begins to glow and we can *see* it's hot!

The problem with these subjective perceptions of heat is that they may not be the same for everybody. If our two hands can't agree on whether water is warm or cold, how likely is it that a group of people can set a uniform standard? We need to construct a device of some kind that responds to temperature in a simple, measurable way—we need a thermometer.

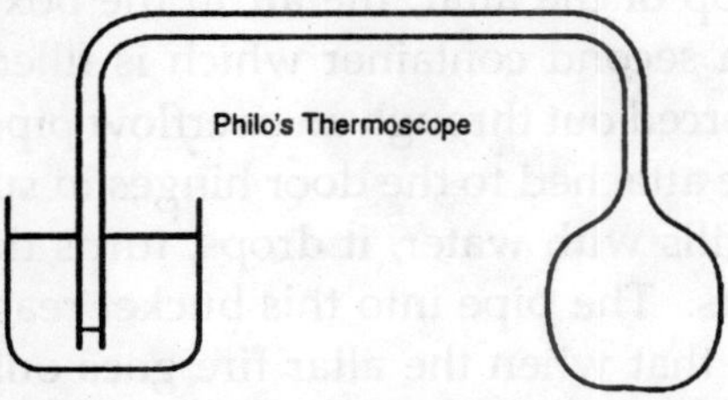

The first step on the road to a thermometer was taken by one Philo of Byzantium, an engineer, in the second century

BC. He took a hollow lead sphere connected with a tight seal to one end of a pipe, the other end of the pipe being under water in another vessel.

To quote Philo: "*...if you expose the sphere to the sun, part of the air enclosed in the tube will pass out when the sphere becomes hot. This will be evident because the air will descend from the tube into the water, agitating it and producing a succession of bubbles.*

Now if the sphere is put back in the shade, that is, where the sun's rays do not reach it, the water will rise and pass through the tube ..."

"No matter how many times you repeat the operation, the same thing will happen.

In fact, if you heat the sphere with fire, or even if you pour hot water over it, the result will be the same."

Notice that Philo did what a real investigative scientist should do—he checked that the experiment was *reproducible,* and he established that the air's expansion was in response to *heat* being applied to the sphere, and was *independent of the source of the heat.*

This expansion of air on heating became widely known in classical times, and was used in various dramatic devices. For example, Hero of Alexandria describes a small temple where a fire on the altar causes the doors to open.

The altar is a large airtight box, with a pipe leading from it to another enclosed container filled with water. When the fire is set on top of the altar, the air in the box heats up and expands into a second container which is filled with water. This water is forced out through an overflow pipe into a bucket hung on a rope attached to the door hinges in such a way that as the bucket fills with water, it drops, turns the hinges, and opens the doors. The pipe into this bucket reaches almost to the bottom, so that when the altar fire goes out, the water is sucked back and the doors close again. Still, none of these ingenious devices is a *thermometer*. There was no attempt (at least none recorded) by Philo or his followers to make a *quantitative* measurement of how hot or cold the sphere was. And the "meter" in thermometer means measurement.

THE FIRST THERMOMETER

Galileo claimed to have invented the first thermometer. Well, actually, he called it a thermoscope, but he *did* try to measure "degrees of heat and cold" according to a colleague, and that qualifies it as a thermometer. (Technically, a *thermoscope* is a device making it possible to *see* a temperature change, a *thermometer* can *measure* the temperature change.) Galileo used an inverted narrow-necked bulb with a tubular neck, like a hen's egg with a long glass tube attached at the tip. He first heated the bulb with his hands then immediately put it into water. He recorded that the water rose in the bulb the height of "one palm". Later, either Galileo or his colleague Santorio Santorio put a paper scale next to the tube to read off changes in the water level. This definitely made it a thermometer, but who thought of it first isn't clear (they argued about it). And, in fact, this thermometer had problems.

Question: what problems? If you occasionally top up the water, why shouldn't this thermometer be good for recording daily changes in temperature?

Answer: because it's also a barometer! But—Galileo didn't know about the atmospheric pressure.

Torricelli, one of Galileo's pupils, was the first to realise, shortly after Galileo died, that the real driving force in suction was external atmospheric pressure, a satisfying mechanical explanation in contrast to the philosophical "nature abhors a vacuum". In the 1640's, Pascal pointed out that the *variability* of atmospheric pressure rendered the air thermometer untrustworthy.

Liquid-in-glass thermometers were used from the 1630's, and they were of course insensitive to barometric pressure. Meteorological records were kept from this time, but there was no real uniformity of temperature measurement until Fahrenheit, almost a hundred years later.

NEWTON'S ANONYMOUS TABLE OF TEMPERATURES

The first systematic account of a range of different temperatures, "Degrees of Heat", was written by Newton, but published anonymously, in 1701. Presumably he felt that this project lacked the timeless significance of some of his other achievements. Taking the freezing point of water as zero, Newton found the temperature of boiling water to be almost three times that of the human body, melting lead eight times as great (actually 327C, whereas 8x37=296, so this is pretty good!) but for higher temperatures, such as that of a wood fire, he underestimated considerably. He used a linseed oil liquid in glass thermometer up to the melting point of tin (232°C). Newton tried to estimate the higher temperatures indirectly. He heated up a piece of iron in a fire, then let it cool in a steady breeze. He found that, at least at the lower temperatures where he could cross check with his thermometer, the temperature dropped in a geometric progression, that is, if it took five minutes to drop from 80° above air temperature to 40° above air temperature, it took another five minutes to drop to 20° above air, another five to drop to 10° above, and so on. He then assumed this same pattern of temperature drop was true at the high temperatures beyond the reach of his thermometer, and so estimated the temperature of the fire and of iron glowing red hot. This wasn't very accurate—he (under)estimated the temperature of the fire to be about 600°C.

FAHRENHEIT'S EXCELLENT THERMOMETER

The first really good thermometer, using mercury expanding from a bulb into a capillary tube, was made by Fahrenheit in the early 1720's. He got the idea of using mercury from a colleague's comment that one should correct a *barometer* reading to allow for the variation of the density of mercury with temperature. The point that has to be borne in mind in constructing thermometers, and defining temperature scales, is that not all liquids expand at uniform rates on heating—water, for example, at first contracts on heating from its freezing point, then begins to expand at around forty degrees Fahrenheit, so a water thermometer wouldn't be very helpful on a cold day. It is also not easy to manufacture a *uniform cross section* capillary tube, but Fahrenheit managed to do it, and demonstrated his success by showing his thermometers agreed with each other over a whole range of temperatures. Fortunately, it turns out that mercury is well behaved in that the temperature scale defined by taking its expansion to be uniform coincides very closely with the true temperature scale.

Amontons' Air Thermometer:

A little earlier (1702) Amontons introduced an *air pressure thermometer*. He established that if air at atmospheric pressure (he states 30 inches of mercury) at the freezing point of water is enclosed then heated to the boiling point of water, but meanwhile kept at constant volume by increasing the pressure on it, the pressure goes up by about 10 inches of mercury. He also discovered that if he compressed the air in the first place, so that it was at a pressure of sixty inches of mercury at the temperature of melting ice, then if he raised its temperature to that of boiling water, at the same time adding mercury to the column to keep the volume of air constant, the pressure increased by 20 inches of mercury. In other words, he found that for a fixed amount of air kept in a container at constant volume, the pressure increased with temperature by about 33per cent from freezing to boiling, that percentage being *independent of the initial pressure.*

THE ZEROTH LAW OF THERMODYNAMICS

Once the thermometer came to be widely used, more precise observations of temperature and heat flow became possible. Joseph Black, a professor at the University of Edinburgh in the 1700's, noticed that a collection of objects at different temperatures, if brought together, will all eventually reach the same temperature. As he wrote, "By the use of these instruments [thermometers] we have learned, that if we take 1000, or more, different kinds of matter, such as metals, stones, salts, woods, cork, feathers, wool, water and a variety of other fluids, although they be all at first of different heats, let them be placed together in a room without a fire, and into which the sun does not shine, the heat will be communicated from the hotter of these bodies to the colder, during some hours, perhaps, or the course of a day, at the end of which time, if we apply a thermometer to all of them in succession, it will point to precisely the same degree."

We say nowadays that bodies in "thermal contact" eventually come into "thermal equilibrium"—which means they finally attain the same temperature, after which no further heat flow takes place. This is equivalent to:

If two objects are in thermal equilibrium with a third, then they are in thermal equilibrium with each other.

The "third body" in a practical situation is just the thermometer. It's perhaps worth pointing out that this trivial sounding statement certainly wasn't obvious before the invention of the thermometer. With only the sense of touch to go on, few people would agree that a piece of wool and a bar of metal, both at 0°C, were at the same temperature.

"Flow of heat" that takes place between bodies as they move towards thermal equilibrium? For example, suppose I reproduce one of Fahrenheit's experiments, by taking 100 ccs of water at 100°F, and 100ccs at 150°F, and mix them together in an insulated jug so little heat escapes. What is the final temperature of the mix? Of course, it's close to 125°F—not surprising, but *it does tell us something!* It tells us that the amount of heat required to raise the temperature of 100 cc of water from 100°F to 125°F is *exactly the same* as the amount needed to raise it from 125°F to 150°F. A series of such experiments (done by Fahrenheit, Black and others) established that it *always* took the same amount of heat to raise the temperature of 1 cc of water by one degree, *independent* of the initial temperature of the water (provided it's between the freezing point and the boiling point). This makes it possible to define a unit of heat. Perhaps unfairly to Fahrenheit,

1 calorie is the heat required to raise the temperature of 1 gram of water by 1 degree Celsius.

(Celsius also lived in the early 1700's. His scale has the freezing point of water as 0°C, the boiling point as 100°C. Fahrenheit's scale is no longer used in science, but lives on in engineering in the US, and in the British Thermal Unit, which is the heat required to raise the temperature of one pound of water by 1°F.)

SPECIFIC HEATS AND CALORIMETRY

The specific heat of a substance is the heat required in calories to raise the temperature of 1 gram by 1 degree Celsius.

As Fahrenheit continues his measurements of heat flow, it quickly became evident that for different materials, the amount of heat needed to raise the temperature of one gram by one degree could be quite different. It was natural to measure specific heats relative to that of water, the simplest and most readily available substance for calorimetric experiments. There were some surprises. For example, it had been widely thought before the measurements were made, that one cc of Mercury, being a lot heavier than one cc of water, would take more heat to raise its temperature by one degree. This proved not to be the case—Fahrenheit himself made the measurement. In an insulating container, called a "calorimeter" he added 100ccs of water at 100°F to 100ccs of mercury at 150°F, and stirred so they quickly reached thermal equilibrium.

Question: what do you think the final temperature was?
Answer: The final temperature was, surprisingly, about 120°F. 100 cc of water evidently "contained more heat" than 100 cc of mercury, despite the large difference in weight!

The technique, called calorimetry, was widely used to find specific heats of many different substances, and at first no clear pattern emerged. It was puzzling that the specific heat of mercury was so low compared with water. As the experiments progressed, it gradually became evident that heavier substances, paradoxically, had lower specific heats.

Meanwhile, this quantitative approach to scientific observation had spread to chemistry. Towards the end of the

1700's, Lavoisier weighed chemicals involved in reactions before and after the reaction. This involved weighing the gases involved, so had to be carried out in closed containers, so that, for example, the weight of oxygen used and the carbon dioxide, etc., produced would accounted for in studying combustion. *The big discovery was that mass was neither created nor destroyed.* This had not been realised before because no one had weighed the gases involved. It made the atomic theory suddenly more plausible, with the idea that maybe chemical reactions were just rearrangements of atoms into different combinations.

Lavoisier also clarified the concept of an element, an idea that was taken up in about 1800 by John Dalton, who argues that a given compound consisted of identical molecules, made up of elementary atoms in the same proportion, such as H_2O (although that was thought initially to be HO). This explained why, when substances reacted chemically, such as the burning of hydrogen to form water, it took exactly eight grams of oxygen for each gram of hydrogen. (Well, you could also produce H_2O_2 under the right conditions, with exactly sixteen grams of oxygen to one of hydrogen, but the simple ratios of amounts of oxygen needed for the two reactions were simply explained by different molecular structures, and made the atomic hypothesis even more plausible.) Much effort was expended carefully weighing the constituents in many chemical reactions, and constructing diagrams of the molecules. The important result of all this work was that it became possible to list the relative weights of the atoms involved. For example, the data on H_2O and H_2O_2 led to the conclusion that an oxygen atom weighed sixteen times the weight of a hydrogen atom. It must be emphasized, though, that these results gave no clue as to the actual weights of atoms! All that was known was that atoms were too small to see in the best microscopes. Nevertheless, knowing the relative weights of some atoms in 1820 led to an important discovery. Two professors in France, Dulong and Petit, found that for a whole series of elements *the product of atomic weight and specific heat* was the same.

Element	Specific Heat (cals/gm/ degree C)	Relative weights of the atoms	Product of relative atomic weight and specific heat
Lead	0.0293	12.95	0.3794
Tin	0.0514	7.35	0.3779
Zinc	0.0927	4.03	0.3736
Sulphur	0.1880	2.011	0.3780

The significance of this, as they pointed out, was that the "specific heat", or heat capacity, of each *atom* was the same—a piece of lead and a piece of zinc having the same number of atoms would have the same heat capacity. So heavier atoms absorbed no more heat than lighter atoms for a given rise in temperature. This partially explained why mercury had such a surprisingly low heat capacity. Of course, having no idea how big the atoms might be, they could go no further. And, indeed, many of their colleagues didn't believe in atoms anyway, so it was hard to convince them of the significance of this discovery.

LATENT HEAT

One of Black's experiments was to set a pan of water on a steady fire and observe the temperature as a function of time. He found it steadily increased, reflecting the supply of heat from the fire, until the water began to boil, whereupon the temperature stayed the same for a long time. The steam coming off was at the same (boiling) temperature as the water. So what was happening to the heat being supplied? Black correctly concluded that heat needed to be supplied to change water from its liquid state to its gaseous state, that is, to steam. In fact, a lot of heat had to be supplied: 540 calories per gram, as opposed to the mere 100 calories per gram needed to bring it from the freezing temperature to boiling. He also discovered that it took 80 calories per gram to melt ice into water, with no rise in temperature. This heat is released when the water freezes back to ice, so it is somehow "hidden" in the water. He called it *latent* heat, meaning hidden heat.

Chapter 2

Thermal Expansion and the Gas Law

Coefficients of Expansion

Almost all materials expand on heating—the most famous exception being water, which contracts as it is warmed from 0 degrees Celsius to 4 degrees. This is actually a good thing, because as freezing weather sets in, the coldest water, which is about to freeze, is less dense than slightly warmer water, so rises to the top of a lake and the ice begins to form there. For almost all other liquids, solidification on cooling begins at the bottom of the container. So, since water behaves in this weird way, ice skating is possible! Also, as a matter of fact, life in lakes is possible—the ice layer that forms insulates the rest of the lake water from very cold air, so fish can make it through the winter.

LINEAR EXPANSION

The coefficient of linear expansion α of a given material, for example a bar of copper, at a given temperature is defined as the fractional increase in length that takes place on heating through one degree:

$$L \rightarrow L + \Delta L = (1 + \alpha)L \text{ when } T \rightarrow T + 1°C$$

Of course, α might vary with temperature (it does for water, as we just mentioned) but in fact for most materials it stays close to constant over wide temperature ranges.

For copper, $\alpha = 17 \times 10^{-16}$.

VOLUME EXPANSION

For *liquids and gases,* the natural measure of expansion is the coefficient of volume expansion, β.

$$V \rightarrow V + \Delta V = (1 + \beta)V \text{ when } T \rightarrow T + 1°C$$

Of course, on heating a bar of copper, clearly the *volume* as well as the *length* increases—the bar expands by an equal fraction in all directions (this could be experimentally verified, or you could just imagine a cube of copper, in which case all directions look the same).

The volume of a cube of copper of side L is $V = L^3$. Suppose we heat it through one degree. Putting together the definitions of α, β above,

$$V \rightarrow (1 + \beta)V,\ L \rightarrow (1+\alpha)L,\ L^3 \rightarrow (1+\alpha)^3 L^3 \text{ or } V \rightarrow (1 + \alpha)^3 V.$$

So $(1 + \beta) = (1+\alpha)^3$. But remember α is very, very small—so even though $(1+\alpha)^3 = 1 + 3\alpha + 3\alpha^2 + \alpha^3$, the last two terms are *completely negligible* (check it out!) so to a fantastically good approximation:

$$\beta = 3\alpha.$$

The coefficient of volume expansion is just three times the coefficient of linear expansion.

GAS PRESSURE INCREASE WITH TEMPERATURE

In 1702, Amontons discovered a *linear* increase of P with T for air, and found P to increase about 33per cent from the freezing point of water to the boiling point of water. That is to say, he discovered that if a container of air were to be sealed at 0°C, at ordinary atmospheric pressure of 15 pounds per square inch, and then heated to 100°C but kept at the same volume, the air would now exert a pressure of about 20 pounds per square inch on the sides of the container. (Of course, strictly speaking, the container will also have increased in size, that would lower the effect—but it's a tiny correction, about ½per cent for copper, even less for steel and glass.) Remarkably, Amontons discovered, if the gas were initially at a pressure of *thirty* pounds per square inch at 0°C, on heating to 100°C the pressure would go to about 40 pounds per square inch—so the *percentage* increase in pressure was the same for *any* initial pressure: on heating through 100°C, the pressure would always increase by about 33per cent. Furthermore, the result turned out to be the same for different gases! Finding a Natural Temperature Scale.In class, we plotted air pressure as a function of temperature for a fixed volume of air, by making several measurements as the air was slowly heated (to give it a chance to all be at the same temperature at each stage). We found a straight line. On the graph, we extended the line backwards, to see how the pressure would presumably drop on cooling the air. We found the remarkable prediction that the pressure should drop to zero at a temperature of about –273°C. In fact, if we'd done the cooling experiment, we would have found that air doesn't actually follow the line all the way down, but condenses to a liquid at around -200°C. However, helium gas stays a gas almost to –270°C, and follows the line closely. We shall discuss the physics of gases, and the interpretation of this, much more fully in a couple of lectures. For now, the important point is that this suggests a *much more natural temperature scale* than the Celsius one: we should take –273°C as the zero of temperature! For one thing, if we do that, the pressure/temperature relationship for a gas becomes beautifully simple:

$$P \alpha T.$$

This temperature scale, in which the degrees have the same size as in Celsius, is called the Kelvin or absolute scale. Temperatures are written 300K. To get from Celsius to Kelvin, just add 273.15.

AN IDEAL GAS

Physicists at this point introduce the concept of an "Ideal Gas". This is like the idea of a frictionless surface: it doesn't exist in nature, but it is a very handy approximation to some real systems, and makes problems much easier to handle mathematically. The ideal gas is one for which $P \propto T$ for all temperatures, so helium is close to ideal over a very wide range, and air is close to ideal at ordinary atmospheric temperatures and above.

THE GAS LAW

We say earlier in the course that for a gas at constant *temperature* PV = constant (Boyle's Law). Now at constant *volume,*

$$P \alpha T.$$

We can put these together in one equation to find a relationship between pressure, volume and temperature:

$$PV = CT$$

where C is a constant. Notice, by the way, that we can immediately conclude that at fixed pressure, $V \propto T$, this is called *Charles' Law*. (*Exercise*: prove from this that the coefficient of volume expansion of a gas varies significantly with temperature.)But what is C? Obviously, it depends on how much gas we have—double the amount of gas, keeping the pressure and temperature the same, and the volume will be doubled, so C will be doubled. But notice that C will *not* depend on what gas we are talking about: if we have two separate one-liter containers, one filled with hydrogen, the other with oxygen, both at atmospheric pressure, and both at the same temperature, then C will be the same for both of them. One might conclude from this that C should be defined for one liter of gas at a specified temperature and pressure, such as 0°C and 1 atmosphere, and that could be a consistent scheme. It might seem more natural, though, to specify a particular *mass* of gas, since then we wouldn't have to specify a particular temperature and pressure in the definition of C. But that idea brings up a further problem: one gram of oxygen takes up a lot less room than one gram of hydrogen. Since we've just seen that choosing the same *volume* for the two gases gives the same constant C for the two gases, evidently taking the same *mass* of the two gases will give different C's.

AVOGADRO'S HYPOTHESIS

The resolution to this difficulty is based on a remarkable discovery the chemists made two hundred years or so ago: they found that *one liter* of nitrogen could react with exactly *one liter* of oxygen to produce exactly *two liters* of NO, nitrous oxide, all volume measurements being at the same temperature and pressure. Further, one liter of oxygen combined with *two* liters of hydrogen to produce two liters of steam.

These simple ratios of interacting gases could be understood if one imagined the atoms combining to form molecules, and made the further assumption, known as Avogadro's Hypothesis (1811):

Equal volumes of gases at the same temperature and pressure contain the same number of molecules.

One could then understand the simple volume results by assuming the gases were made of diatomic molecules, H_2, N_2, O_2 and the chemical reactions were just molecular recombinations given by the equations $N_2 + O_2 = 2NO$, $2H_2 + O_2 = 2H_2O$, etc. Of course, in 1811 Avogadro didn't have the slightest idea what this number of molecules was for, say, one liter, and nobody else did either, for another fifty years. So no-one knew what an atom or molecule weighed, *but* assuming that chemical reactions were atoms combining into molecules, or rearranging from one molecular pairing or grouping to another, they could figure out the *relative* weights of atoms, such as *an oxygen atom had mass 16 times that of a hydrogen atom*— even though they had no idea how big these masses were!

This observation led to defining *the natural mass of a gas* for setting the value of the constant C in the gas law to be *a "mole" of gas*: hydrogen was known to be H_2 molecules, so *a mole of hydrogen was 2 grams,* oxygen was O_2, so a mole of oxygen was 32 grams, and so on. With this definition, a mole of oxygen contains the same number of molecules as a mole of hydrogen: so at the same temperature and pressure, they will occupy the same volume. At 0°C, and atmospheric pressure, the volume is 22.4 liters. So, for one mole of a gas (for example, two grams of hydrogen), we set the constant C equal to R, known as the universal gas constant, and equal to 8.3 J/(mol.K) and $PV = RT$.

For n moles of a gas, such as $2n$ grams of hydrogen, the law is:

$$PV = nRT$$

and this is the standard form of the Gas Law.

By the late 1700's, the experiments of Fahrenheit, Black and others had established a systematic, quantitative way of measuring temperatures, heat flows and heat capacities—but this didn't really throw any new light on just *what* was flowing. This was a time when the study of electricity was all the rage, led in America by Benjamin Franklin, who had suggested in 1747 that electricity was one (invisible) fluid (it had previously been suggested that there were two fluids, corresponding to the two kinds of electrical charging observed).

LAVOISIER'S CALORIC FLUID THEORY

Perhaps heat was another of these invisible fluids? In 1787, Lavoisier, the French founder of modern chemistry, thought so, and called it the *caloric fluid*, from the Greek word for heat. (Lavoisier was the first to attempt to list a table of *elements*, to replace the ancient elements of earth, air, water and fire. His list of thirty-three elements included hydrogen, oxygen, sulphur, charcoal, etc., but he also *included caloric*—and light.) The existence of such a fluid was really quite plausible—heat flowed from a hot body to a cold body, and the recent quantitative calorimetric experiments of Black and others seemed to establish that heat was a *conserved quantity*, as one would expect of a fluid. One could also understand some of the well-known effects of heat in terms of a fluid, and establish some of the fluid's properties. For example, since it tended to flow from hot bodies into cold bodies and spread throughout the body, presumably its particles repelled each other, just like those of the electrical fluid. However, in contrast to electricity, which had no noticeable effect on the appearance of a charged object, when heat was added to a solid things changed considerably. First the material expanded, then it changed to a liquid and finally to a gas, if sufficient heat could be delivered. Further heating expanded the gas, or increased its pressure if it was held in a fixed container. To interpret this sequence of events in terms of a caloric fluid being fed into the material, one could imagine the fluid flowing between the atoms of the solid and lessening their attraction for each other, until the solid melted into a liquid, whereupon the caloric continued to accumulate around the atoms until they were pushed apart into a gas. It was thought that in the gas each atom or molecule was surrounded by a ball of caloric, like a springy ball of wool, and these balls were packed in a container like oranges in a crate, except that the caloric balls could expand indefinitely as heat was poured in.Various other effects could be explained by the caloric theory: when a gas is suddenly compressed, it gets hotter because the same amount of caloric is now occupying a smaller volume. When two solids are rubbed together, some caloric is squeezed out at the surfaces, or perhaps tiny pieces of material are rubbed off, and lose their caloric, so heat appears. Radiant heat was presumed to be caloric particles flying through space. Recall that at that time (just before 1800) it was generally accepted that *light* was a stream of particles.

In 1769 a Lancashire wigmaker, Richard Arkwright, patented a successful cotton-spinning machine. Lancashire had been for a long time a center of the textile trade, but before Arkwright the fabrics were woven on hand looms by skilled weavers. The new machines could be operated by less skilled workers, and in fact were largely operated by children, although, in contrast to some of his competitors, Arkwright refused to employ any child younger than six. The motive power driving the machines was at first horses, but in 1771 Arkwright built a large factory containing many machines all driven by a water wheel. This was the beginning of the modern system of mass production. Prices fell, and the skilled hand weavers became impoverished. Our interest in this, however, is not the social consequences, but just the water wheel. Previously, water wheels had been used for centuries to grind flour, and for other purposes, but their efficiency had not been a major concern. In the factory, though, the more efficient the wheel, the more children could be spinning the cotton, and the bigger the profits. Twenty years earlier, John Smeaton (the first Englishman to call himself a civil engineer) had investigated different types of water wheels, and found the overshot type (in which the water pours on to the top of the wheel) to perform best.

The power output of a water wheel can be measured by using it to raise a load—in those days, it would be how many pounds could be raised through one foot per second, say (we would now just use watts, and it's amusing to note that the first unit of power, the horsepower, was proposed in 1783 by James Watt to be 33,000 foot pounds per minute). The ultimate in efficiency would be a *reversible* water wheel, which could be run backwards, to raise the water back again. This is best visualized by having a wheel with a series of buckets attached. Suppose the wheel is run for some time and its power output is used to lift a weight a given distance. Now reverse it, let the weight fall, running the wheel backwards, making sure the buckets now fill at the bottom and empty at the top. How much water is lifted back up? A truly reversible wheel would put all the water back. We know this isn't going to happen, but if the reversed wheel manages to lift half the water back, say, then it's 50per cent efficient.

In building the first factory, the water wheel was not just placed under a waterfall. The water was channeled to it for maximum efficiency. Smeaton had established that the flow of water into the buckets must be *as smooth as possible*. Turbulence was wasted effort—it didn't help the wheel go round. The water should flow onto the wheel, not fall from some height. Finally, the perfect wheel (not quite realizable in practice) would be reversible—it could be run backwards to put the water back up using the same amount of work it delivered in the first place.

CARNOT'S CALORIC WATER WHEEL

Arkwright's factory was so successful that within a few years similar factories had been built wherever a water wheel could be operated economically in Northern England. The next step was to use steam power. As steam engine design improved, the English economy mushroomed far ahead of European competitors—but in contrast to the present day, these technological advances owed virtually nothing to basic science. It was all inspired tinkering.

The first attempt to analyse the steam engine in a scientific way was by a Frenchman, Sadi Carnot, in 1820—and he relied heavily an analogy with the water wheel. In the steam engine, heat is delivered to water which becomes steam and drives a piston. The piston does work, usually by turning a wheel, then waste heat is expelled. Remember that at that time heat was seen as an invisible fluid, impelled by its nature to flow from hot objects to cold objects. Carnot saw this as parallel to water which always flows from high places to low places. A water wheel placed under a waterfall will extract useful work from the falling water, work that otherwise would be wasted. Carnot concluded that the steam engine extracted work from the caloric fluid in a similar way, work which would have been lost if the heat had simply been allowed to flow from a hot object to a cold object.

By analogy with the water wheel, Carnot proved that if all friction were to be eliminated, and heat flow into and out of the steam were isothermal (this is the same as the water being fed onto the wheel at just the right height, so it doesn't fall and waste energy) then one could imagine a reversible engine. As we shall discuss in detail later, he was able to find for such an engine just how much work the engine could perform for a given heat input, and the answer was surprisingly low. Furthermore, no engine could ever be more efficient than a reversible engine, because if it were, it could be used to drive the reversible engine backwards, replacing the heat in the furnace, with energy to spare, and would be a perpetual motion machine.

Carnot's basic assumption that heat is a fluid was flawed, but his reasoning was of sufficient generality that his conclusions about efficiency were correct, and proved to be a crucial step toward understanding engines.

COUNT RUMFORD

The first real attack on the caloric theory of heat took place in a cannon factory in Bavaria, under the direction of one Count Rumford of the Holy Roman Empire. This Count was actually born Benjamin Thompson in Woburn, in the English colony of Massachusetts, in 1753, which he left in a hurry after choosing the wrong side in the Revolutionary War. He was a brilliant man, extraordinarily inventive as a scientist and engineer—but it is difficult to form a coherent picture of his character. He seemed genuinely upset by the plight of the poor in Munich and made great personal efforts for years to ensure they were properly fed and clothed. Throughout his life, he invented practical devices to make daily living better: stoves, fitted kitchens, drip coffeepots, lighting, and many more. Yet, despite this love for humanity and his clear desire to make life better for everyone, Rumford did not apparently like—or get on with—actual people. The only exceptions were those with power who might prove useful, and almost any attractive woman he met. Rumford dumped his own family unceremoniously when war broke out and he fled to England. When garrisoned on Long Island in 1782 (fighting for the British) he treated the local people horribly. He always engaged in shameless self-promotion, often with little regard for the truth. But he *did* make important contributions to many fields: food, clothing, work and education for the poor both in Bavaria and (less successfully) later in England, and all manner of engineering improvements, from the domestic devices listed above to state of the art artillery. In fact, his artillery designs were so highly regarded that by 1799 US President Adams tried to persuade him to return to America to found a Military Academy, with assurances that all was forgiven.

His father died when Benjamin was still a child, and although his mother remarried, he felt strongly that he had to take care of himself. He worked hard at school, then at age eighteen began working as a tutor for children of rich families, and after a short time became a teacher in a school in Concord, New Hampshire. At nineteen, he married a rich young widow, who decided to upgrade his appearance to fit in better with her friends. She bought him a scarlet hussar cloak, they used a two-horse chaise called a curricle, the only other curricle in the province belonging to the Royal Governor, John Wentworth. The marriage took place one day after the Governor had reviewed the troops, and the bride and groom were guests at the Governor's table. Thompson assiduously cultivated the Governor. They went together on a surveying expedition exploring the hilly country of the province. Thompson's real ambition was military, and in 1774 Governor Wentworth appointed him a major in the New Hampshire Militia. That year, the people were becoming increasingly rebellious against British rule and British taxes. Order was kept, at least in part, by the British Army. Thompson was part of a scheme to discourage desertion from that army, and when this became known to those plotting revolution, they declared him a "Rebel to the State". He moved rapidly to London, abandoning his wife and two-month-old daughter. In London, he supplied military intelligence, and was rewarded with a salaried position involving no work.

Ever the scientist (with a military bent), he spent a lot of time on gunnery experiments. He used a ballistic pendulum to find how the speed of a bullet was affected by small changes in gun design and in the gunpowder mix. He disproved the widely held view that slightly damp gunpowder was actually more effective.

He made a trip back to America in 1781. He wintered with a few hundred soldiers at Huntington, Long Island, in 1782, setting up camp in the churchyard, and forcing the local people to build fortifications for his camp using the church timbers and their rail fences. He constructed bread ovens using the tombstones. He designed a new gun carriage that could be disassembled, carried by three horses, then put together and fired in a minute and a quarter. But the war was over. He burned all the wood, rather than give it back to the people, and returned to London.

Still enthusiastic for military adventure, he decided to go to Vienna, which seemed a likely trouble spot. He first got himself promoted to full colonel, acquired the splendid uniform, and had his portrait painted by Thomas Gainsborough. When he reached Strasbourg, it happened that a military review was underway. Naturally, Thompson appeared in full regalia, impressing the reviewer, who was a nephew of the Elector of Bavaria. This landed him an important post in Munich, the capital of Bavaria. Among his other duties, he was to organize the feeding and clothing of the army. He took a practical, scientific approach. He had each garrison maintain its own vegetable garden, and gave directions on how to use them most effectively from a nutritional point of view. He ran experiments on the relative thermal conductivity of various fabrics, and found that trapped air in fabrics was the most important measure of heat insulation. He decided that uniforms should be cotton in summer, wool in winter. He invented thermal underwear. In 1792, he became Count Rumford.

It should also be mentioned that he greatly improved the city in many ways: he invented the soup kitchen for the poor, an idea which spread throughout Europe. In Switzerland, the meal tickets had Rumford's picture on them. He built workhouses, to provide work for the unemployed in making uniforms for the military. He helped design and lay out a beautiful park, where a memorial to him still stands. He also worked on many domestic improvements, such as the Rumford stove for more heat and less smoke, coffee makers, and an efficient but soft light, this last at least partly because, to quote him, "that mysterious light which comes from bodies moderately illuminated is certainly most favourable to female beauty" and Rumford was a great connoisseur of female beauty.

RUMFORD'S THEORY OF HEAT

The contribution to physics for which he is most remembered took place in Munich, and he stumbled into it more or less accidentally. But, as he remarks in presenting his findings to the Royal Society in 1798, "a habit of keeping the eyes open to everything that is going on in the ordinary course of the business of life has oftener led, as it were by accident, ... to sensible schemes for investigation ... than all the more intense meditations of philosophers, in the hours expressly set apart for study." What he was looking at was cannon boring, beefing up the Bavarian artillery in case of attack by the French, but what he was thinking about was whether or not Lavoisier's calorific fluid really existed. He was skeptical. Cannon were bored by turning an iron bit inside a brass cylinder, the power being supplied by horses. The friction of the iron bit on the brass generated heat. This was accounted for in the caloric theory by the pressure and movement squeezing out caloric fluid, in particular from the fragments that were sheared off. Rumford carefully collected these fragments and found them to be identical to the ordinary metal in heat capacity, etc., they didn't seem to have lost anything. Then he measured the heat production for an extended period, by having the brass cylinder immersed in water, and insulated. After extended grinding, the water (two gallons) began to boil. This was a startling event! To quote from his account:

"At 2 hours and 20 minutes it was 200°; and at 2 hours 30 minutes it Actually Boiled!

It would be difficult to describe the surprise and astonishment expressed in the countenances of the by-standers, on seeing so large a quantity of cold water heated, and actually made to boil without any fire."

Rumford goes on the analyse the whole experiment quantitatively: he gives the weight of the box, and so estimates how much heat it absorbs, as well as other parts of the apparatus which became warm, and measures the rate of cooling with the grinding stopped to estimate how much heat leaked out during the run. Taking all these factors into account, he estimated that heat production was equivalent to nine ¾ inch candles burning continuously. Long before the concept was formulated, Rumford had measures the mechanical equivalent of heat, at least approximately. In fact, many years later, Joule went over his Rumford's figures and found he was within 20per cent or so of the right answer. Rumford realised, of course, this wasn't a good way to produce heat—as he remarked, more heat could have been gained simply by burning the horses' fodder. His real interest here was in demolishing the caloric theory.

He concluded:...we must not forget to consider that most remarkable circumstance, that the source of the Heat generated by Friction, in these Experiments, appeared evidently to be *inexhaustible.* It is hardly necessary to add, that anything which any insulated body, or system of bodies, can continue to furnish without limitation, cannot possibly be a material substance: and it appears to me to be extremely difficult, if not quite impossible, to form any distinct idea of any thing, capable of being excited and communicated, in the manner the heat was excited and communicated in these Experiments, except it be Motion.

Exactly what Rumford meant by Motion has been debated, but it was some type of internal vibration of material, perhaps only distantly related to our modern, atom based, picture of heat vibrations. Still, by casting real doubt on the caloric theory, it was a step in the right direction. He had also established that if a caloric fluid really existed, it was certainly very light! He took three identical glass bottles containing equal weights of water, mercury and alcohol respectively, made them exactly equal in weight by tying small lengths of wire around the necks, then cooled them until the water froze, and weighed them again. The latent heat of freezing, and the very different heat capacities of the three fluids, should have resulted in quite different amounts of caloric fluid leaving the three bottles, yet their weights remained exactly the same, within one part in a million (the claimed accuracy of the balance).

After he returned to London in 1798, Rumford planned to repeat some of the public welfare successes in Munich. He wanted to build soup kitchens and workhouses for the poor. He also planned to found an institution which would not only facilitate the implementation of new scientific discoveries in improving living standards, but also train working class men to become mechanics. This became the Royal Institution. Unfortunately, Rumford was difficult to work with, and he did not see eye to eye with the first director, a young Cornishman called Humphry Davy. Sad to report, Rumford's idealistic notions for schooling the poor and improving living standards did not become a priority for the Institution, except for a series of lectures for the public which evolved into entertainments for the wealthy. Nevertheless, the Institution did maintain a first class laboratory in which Davy discovered new elements, including sodium and potassium, and has in fact been an excellent center of scientific research for the last two hundred years.

Chapter 3

The Discovery of Energy Conservation

ROBERT MAYER AND THE COLOUR OF BLOOD

Julius Robert Mayer was born in the mill town of Heilbronn, Germany, on the river Neckar, in 1814. The town's whole economy was based on water power. The ten year old Mayer had a great idea: why not use part of a water wheel's output to drive an Archimedean screw which would pump the water back up again? That way you wouldn't have to rely on the river at all! He decided to build a model. His first try didn't work—some water was pumped back up, but not enough. But surely that could be taken care of by putting in a gear train to make the screw run faster? Disappointingly, he found the water wheel had a tougher time turning the screw faster, and he needed to supply a lot more water over the wheel, so he was back to square one. Increasingly ingenious but unsuccessful fixes finally convinced him that in fact there was no solution: *there was no way to arrange a machine to do work for nothing*. This lesson stayed with Mayer for life.

Mayer studied to become a medical doctor (his studies included one physics course) and in 1840, at age 25, he signed on as a ship's doctor on a ship bound for the tropics. Shortly after reaching the Dutch East Indies, some of the sailors became ill and Mayer's treatment included blood letting. He was amazed to find that the venous blood was a bright red, almost the same as arterial blood. Back in Germany, the venous blood was much darker, and there was a reason: the chemist Lavoisier had explained that the body's use of food, at least in part, amounted to burning it in a controlled way to supply warmth. The darker venous blood in effect contained the ashes, to be delivered to the lungs and expelled as carbon dioxide. Mayer concluded that less burning of food was needed to keep warm in the tropics, hence the less dark blood.

Now, Lavoisier had claimed that the amount of heat generated by burning, or oxygenation, of a certain amount of carbon did not depend on the sequence of chemical reactions involved, so the heat generated by blood chemistry oxygenation would be the same as that from uncontrolled old-fashioned burning in air. This quantitative formulation led Mayer to think about how he would measure the heat generated in the body, to equate it to the food burned. But this soon led to a problem. Anyone can generate extra heat, just by rubbing the hands together, or, for example, by turning a rusty, unoiled wheel: the axle will get hot. Does this 'outside' heat also count as generated by the food? Presumably yes, the food powers the body, and the body generates the heat, even if indirectly. Mayer was convinced from his childhood experience with the water wheel that nothing came from nothing: that outside heat could not just appear from nowhere, it had to have a cause. But he saw that if the indirectly generated heat must also be included, there is a problem. His analysis ran something like this (I've changed the illustration slightly, but the idea's the same): suppose two people are each steadily turning large wheels at the same rate, and the wheels are equally hard to turn. One of them is our rusty unoiled wheel from the last paragraph, and all that person's efforts are going into generating heat. But the other wheel has a smooth, oiled axle and generates a negligible amount of heat. It is equally hard to turn, though, because it is raising a large bucket of water from a deep well. How do we balance the 'food = heat' budget in this second case?

Mayer was forced to the conclusion that for the 'food = heat' equation to make sense, there had to be *another* equivalence: a certain amount of mechanical work, measured for example by raising a known weight through a given distance, had to count the same as a certain amount of heat, measured by raising the temperature of a fixed amount of water, say, a certain number of degrees. In modern terms, a joule has to be equivalent to a fixed number of calories. Mayer was the first to spell out this 'mechanical equivalent of heat' and in 1842 he calculated the number using results of experiments done earlier in France on the specific heats of gases French experimenters had measured the specific heat of the same gas at constant pressure (C_v) and at constant pressure (C_p). They always found C_p to be greater than C_v. Mayer interpreted this with the following thought experiment: consider two identical vertical cylinders, closed at the top by moveable pistons, the pistons resting on the gas pressure, each enclosing the same amount of the same gas at the same temperature. Now supply heat to the two gases, for one gas keep the piston fixed, for the other allow it to rise. Measure how much heat is needed to raise the gas temperature by ten degrees, say. It is found that extra heat is needed for the gas at constant pressure, the one where the piston was allowed to rise. Mayer asserted this was because in that case, some of the heat had been expended as work to raise the piston: this followed very naturally from his previous thinking, and the French measurements led to a numerical value for the equivalence. Mayer understood the sequence: a chemical reaction produces heat and work, that work can then produce a definite amount of heat. This amounted to a statement of the conservation of energy. Sad to report, Mayer was not part of the German scientific establishment, and this ground-breaking work was ignored for some years.

JAMES JOULE

Meanwhile, in Manchester, England, the center of the industrial revolution, the same problem was being approached from quite a different direction by James Joule, the son of a prosperous brewer. Joule was lucky in that as a teenager, he was tutored at home, along with his brother, by John Dalton, the chemist who founded the atomic theory. Manchester was at the cutting edge of technological progress, and one exciting idea in the 1830's was that perhaps coal-driven steam engines could be replaced by battery-driven electric motors. Joule, in his twenties, set himself the task of improving the electric motor to the point where it would be competitive with the steam engine. But it was not to be—after years of effort, he concluded that at best it would take five pounds of zinc consumed in a battery to deliver the work from one pound of coal. But he learned a lot. He found an electric current in a wire produced heat at a rate I^2R, now known as Joule heating. The caloric theory interpretation was that caloric fluid originally in the battery was released along with the electric current and settled in the wire. However, Joule discovered the same heating took place with a current generated by moving the wire past a permanent magnet. It was difficult to see how the caloric fluid got into the wire in that situation. Joule decided the caloric theory was suspect. He generated a current by applying a measured force to a dynamo, and established that the heat appearing in the wire was always directly proportional to the work done by the force driving the dynamo.

Finally, it dawned on him that the electrical intermediary was unnecessary: the heat could be produced directly by the force, if instead of turning a dynamo, it turned paddle wheels churning water in an insulated can.

The paddle wheels turn through holes cut in stationary brass sheets, churning up the water. This is all inside an insulated can, of course. In this way, Joule measured the mechanical equivalent of heat, the same number Mayer had deduced from the French gas experiments. Joule's initial reception by the scientific establishment was not too different from Mayer's. He, too, was a provincial, with a strange accent. But he had a lucky break in 1847, when he reported his work to a meeting of the British Association, and William Thomson was in the audience. Thomson had just spent a year in Paris. He was fully familiar with Carnot's work, and believed the caloric theory to be correct. But he knew that if Joule really had produced heat by stirring water, the caloric theory must be wrong—he said there were 'insuperable difficulties' in reconciling the two.

Mayer and Joule, using entirely different approaches, arrived almost simultaneously at the conclusion that heat and mechanical work were numerically equivalent: a given amount of work could be transformed into a quantitatively predictable amount of heat. Which of the two men deserves more credit (not to mention other contenders!) has been argued for well over a century. Briefly, it is generally conceded that Mayer was the first to spell out the concept of the mechanical equivalent of heat (although closely followed, independently, by Joule) and Joule was the first to put it on a firm experimental basis.

In fact, by the 1840's, although many still believed in the caloric theory, it had run into other difficulties. Before the 1820's, almost everyone believed, following Newton, that light was a stream of particles. Around 1800, Herschel discovered that on passing sunlight through a prism, and detecting the heat corresponding to the different colours, in fact there was heat transmitted beyond the red. This suggested that radiant heat was caloric particles streaming through space, and no doubt very similar in character to light. But in the 1820's it was unambiguously established that light was really a wave. Did this mean heat was a wave too? Perhaps the caloric fluid was oscillations in the ether. Things were now very confused. In 1841, Joule wrote diplomatically: 'let the space between these compound atoms be supposed to be filled with calorific ether in a state of vibration, or, otherwise, to be occupied with the oscillations of the atoms themselves. It transpired, though, that the difficulties in reconciling Carnot's theory and Joule's experiments were not as insuperable as Thomson had claimed. In 1850, a German professor, Rudolph Clausius, pointed out that Carnot's theory was still almost right: the only adjustment needed was that there was a little less heat emerging from the bottom of the 'caloric water wheel' than went in at the top—some of the heat became mechanical energy, the work the steam engine was performing. For real steam engines, the efficiency—the fraction of ingoing heat delivered as useful work—was so low that it was easy to understand why Carnot's picture had been accepted for so long. For the first time, with Clausius' paper, a coherent theory of heat emerged, and the days of the caloric theory drew to a close.

Chapter 4

Kinetic Theory of Gases

BERNOULLI'S PICTURE

Daniel Bernoulli, in 1738, was the first to understand air pressure from a molecular point of view. He drew a picture of a vertical cylinder, closed at the bottom, with a piston at the top, the piston having a weight on it, both piston and weight being supported by the air pressure inside the cylinder. He described what went on inside the cylinder as follows: "let the cavity contain very minute corpuscles, which are driven hither and thither with a very rapid motion; so that these corpuscles, when they strike against the piston and sustain it by their repeated impacts, form an elastic fluid which will expand of itself if the weight is removed or diminished..."Sad to report, his insight, although essentially correct, was not widely accepted. Most scientists believed that the molecules in a gas stayed more or less in place, repelling each other from a distance, held somehow in the ether. Newton had shown that PV = constant followed if the repulsion were inverse-square. In fact, in the 1820's an Englishman, John Herapath, derived the relationship between pressure and molecular speed given below, and tried to get it published by the Royal Society. It was rejected by the president, Humphry Davy, who pointed out that equating temperature with motion, as Herapath did, implied that there would be an absolute zero of temperature, an idea Davy was reluctant to accept. And it should be added that no-one had the slightest idea how big atoms and molecules were, although Avogadro had conjectured that equal volumes of different gases at the same temperature and pressure contained equal numbers of molecules—his famous number—neither he nor anyone else knew what that number was, only that it was pretty big.

MOLECULAR ENERGY AND PRESSURE

It is not difficult to extend Bernoulli's picture to a *quantitative* description, relating the gas pressure to the molecular velocities. As a warm up exercise, let us consider a single perfectly elastic particle, of mass m, bouncing rapidly back and forth at speed v inside a narrow cylinder of length L with a piston at one end, so all motion is along the same line. What is the force on the piston?

Obviously, the piston doesn't feel a smooth continuous force, but a series of equally spaced impacts. However, if the piston is much heavier than the particle, this will have the same effect as a smooth force over times long compared with the interval between impacts. So what is the value of the equivalent smooth force?

Using Newton's law in the form *force = rate of change of momentum,* we see that the particle's momentum changes by $2mv$ each time it hits the piston. The time between hits is $2L/v$, so the frequency of hits is $v/2L$ per second. This means that if there were no balancing force, by conservation of momentum the particle would cause the momentum of the piston to change by $2mv \times v/2L$ units in each second. This is the rate of change of momentum, and so must be equal to the balancing force, which is therefore

$$F = mv^2/L$$

We now generalize to the case of many particles bouncing around inside a rectangular box, of length L in the x-direction (which is along an edge of the box). The total force on the side of area A perpendicular to the x-direction is just a sum of single particle terms, the relevant velocity being the component of the velocity in the x-direction. The *pressure* is just the force per unit area, $P = F/A$. Of course, we don't know what the velocities of the particles are in an actual gas, but it turns out that we don't need the details. If we sum N contributions, one from each particle in the box, each contribution proportional to v_x^2 for that particle, the sum just gives us N times the *average* value of v_x^2. That is to say,

$$P = F/A = Nm\overline{v_x^2}/LA = Nm\overline{v_x^2}/V$$

where there are N particles in a box of volume V. Next we note that the particles are equally likely to be moving in any direction, so the average value of v_x^2 must be the same as that of v_y^2 or v_z^2, and since $v^2 = v_x^2 + v_y^2 + v_z^2$, it follows that

$$P = Nm\overline{v^2}/3V\ .$$

This is a surprisingly simple result! *The macroscopic pressure of a gas relates directly to the average kinetic energy per molecule.* Of course, in the above we have not thought about possible complications caused by interactions between particles, but in fact for gases like air at room temperature these interactions are very small. Furthermore, it is well established experimentally that most gases satisfy the Gas Law over a wide temperature range:

$$PV = nRT$$

for n moles of gas, that is, $n = N/N_A$, with N_A Avogadro's number and R the gas constant.

Introducing Boltzmann's constant $k = R/N_A$, it is easy to check from our result for the pressure and the ideal gas law that *the average molecular kinetic energy is proportional to the absolute temperature,*

$$\overline{E_k} = \overline{\frac{1}{2}mv^2} = \frac{3}{2}kT.$$

Boltzmann's constant $k = 1.38.10^{-23}$ joules/K.

MAXWELL FINDS THE VELOCITY DISTRIBUTION

By the 1850's, various difficulties with the existing theories of heat, such as the caloric theory, caused some rethinking, and people took another look at the kinetic theory of Bernoulli, but little real progress was made until Maxwell attacked the problem in 1859. Maxwell worked with Bernoulli's picture, that the atoms or molecules in a gas were perfectly elastic particles, obeying Newton's laws, bouncing off each other (and the sides of the container) with straight-line trajectories in between collisions. (Actually, there is some inelasticity in the collisions with the sides—the bouncing molecule can excite or deexcite vibrations in the wall, this is how the gas and container come to thermal equilibrium.) Maxwell realised that it was completely hopeless to try to analyse this system using Newton's laws, even though it could be done in principle, there were far too many variables to begin writing down equations. On the other hand, a completely detailed description of how each molecule moved was not really needed anyway. What *was* needed was some understanding of how this microscopic picture connected with the macroscopic properties, which represented averages over huge numbers of molecules.

The *relevant* microscopic information is not knowledge of the position and velocity of every molecule at every instant of time, but just the *distribution function,* that is to say, what percentage of the molecules are in a certain part of the container, and what percentage have velocities within a certain range, at each instant of time. For a gas *in thermal equilibrium,* the distribution function is *independent of time.* Ignoring tiny corrections for gravity, the gas will be distributed uniformly in the container, so the only unknown is the *velocity* distribution function.

VELOCITY SPACE

What does a velocity distribution function look like? Suppose at some instant in time one particular molecule has velocity $\overline{v} = (v_x, v_y, v_z)$. We can record this information by constructing a three-dimensional *velocity space,* with axes v_x, v_y, v_z, and putting in a point P_1 representing the molecule's velocity (the red arrow is of course $\vec{v}$):

Now imagine that at that instant we could measure the velocities of *all* the molecules in a container, and put points $P_2, P_3, P_4, \ldots P_N$ in the velocity space. Since N is of order 10^{21}

for 100 ccs of gas, this is not very practical! But we can imagine what the result would be: a cloud of points in velocity space, equally spread in all directions (there's no reason molecules would prefer to be moving in the *x*-direction, say, rather than the *y*-direction) and thinning out on going away from the origin towards higher and higher velocities.

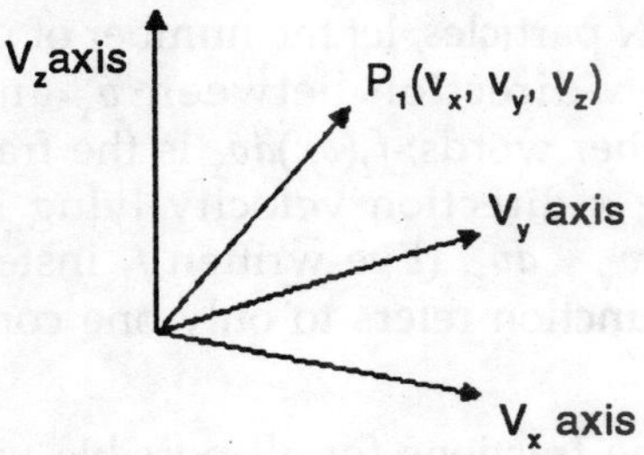

Point P1 represents the locatino of one molecule in velocity space

Now, if we could keep monitoring the situation as time passes individual points would move around, as molecules bounced off the walls, or each other, so you might think the cloud would shift around a bit. But there's a *vast number* of molecules in any realistic macroscopic situation, and for any reasonably sized container it's safe to assume that the number of molecules in any small region of velocity space remains pretty much constant. Obviously, this cannot be true for a region of velocity space so tiny that it only contains one or two molecules on average. But it can be shown statistically that if there are N molecules in a particular small volume of velocity space, the fluctuation of the number with time is of order $\sqrt{N}$, so a region containing a million molecules will vary in numbers by about one part in a thousand, a trillion molecule region by one part in a million. Since 100 ccs of air contains of order 10^{21} molecules, we can in practice divide the region of velocity space occupied by the gas into a billion cells, and *still* have variation in each cell of order one part in a million!

The bottom line is that for a macroscopic amount of gas, fluctuations in density, both in ordinary space and in position space, are for all practical purposes negligible, and we can take the gas to be *smoothly distributed* in both spaces.

MAXWELL'S SYMMETRY ARGUMENT

Maxwell found the velocity distribution function for gas molecules in thermal equilibrium by the following elegant argument based on symmetry.

For a gas of N particles, let the number of particles having velocity in the x-direction between v_x and $v_x + dv_x$ be $Nf_1(vx)dv_x$. In other words, $f_1(vx)dv_x$ is the fraction of all the particles having x-direction velocity lying in the interval between v_x and $v_x + dv_x$. (I've written f_1 instead of f to help remember this function refers to only one component of the velocity vector.)

If we add the fractions for all possible values of v_x, the result must of course be 1:

$$\int_{-\infty}^{\infty} f_1(v_x)dv_x = 1.$$

But there's nothing special about the x-direction—for gas molecules in a container, at least away from the walls, all directions look the same, so the same function f will give the probability distributions in the other directions too. It follows immediately that the probability for the velocity to lie between v_x and $v_x + dv_x$, v_y and $v_y + dv_y$, and v_z and $v_z + dv_z$ must be:

$$Nf_1(v_x)dv_xf_1(v_y)dv_yf_1(v_z)dv_z = nf_1(v_x)f_1(v_y)f_1(v_z)dv_xdv_ydv_z$$

Note that this distribution function, when integrated over all possible values of the three components of velocity, gives the total number of particles to be N, as it should (since integrating over each $f_1(v)dv$ gives unity).

Next comes the clever part—since any direction is as good as any other direction, the distribution function must depend *only* on the total speed of the particle, *not* on the separate velocity components. Therefore, Maxwell argued, it *must* be that:

$$f_1(v_x)f_1(v_y)f_1(v_z) = F(v_x^2 + v_y^2 + v_z^2)$$

where F is another unknown function. However, it is apparent that the *product* of the functions on the left is reflected in the *sum* of variables on the right. It will only come out that way *if the variables appear in an exponent* in the functions on the left. In fact, it is easy to check that this equation is solved by a function of the form:

$$f_1(v_x) = Ae^{-Bv_x^2}.$$

This curve is called a *Gaussian*: it's centered at the origin, and falls off *very* rapidly as v_x increases. Taking $A = B = 1$ just to see the shape, we find:

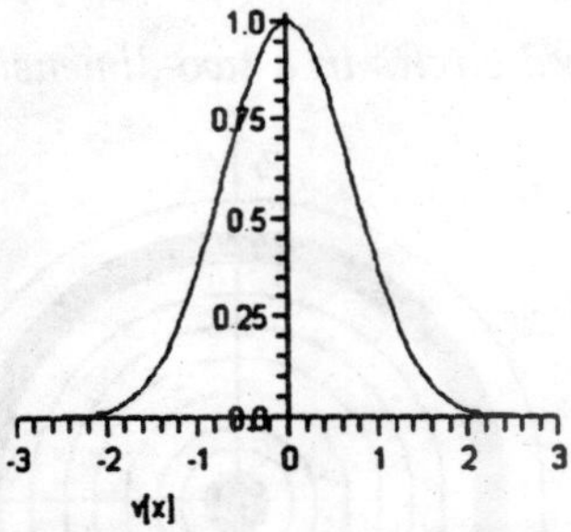

At this point, A and B are arbitrary constants—we shall eventually find their values for an actual sample of gas at a given temperature. Notice that (following Maxwell) we have put a minus sign in the exponent because there must eventually be fewer and fewer particles on going to higher speeds, certainly not a diverging number. Multiplying together the probability distributions for the three directions gives the distribution in terms of particle speed v, where $v^2 = v_x^2 + v_y^2 + v_z^2$. Since all velocity directions are equally likely, it is clear that the natural distribution function is that giving the number of particles having *speed* between v and $v + dv$.From the graph above, it is clear that the most likely value of v_x is zero. If the gas molecules were restricted to one dimension, just moving back and forth on a line, then the most likely value of their *speed* would *also* be zero. However, for gas molecules free to move in two or three dimensions, the most likely value of the speed is *not* zero. It's easiest to see this in a two-dimensional example. Suppose we plot the points P

representing the velocities of molecules in a region near the origin, so the density of points doesn't vary much over the extent of our plot (we're staying near the top of the peak in the one-dimensional curve shown above).Now divide the two-dimensional space into regions corresponding to equal increments in speed:

0 to Δv, Δv to 2Δv, 2Δv to 3Δv,....

In the two-dimensional space, $v = \sqrt{v_x^2 + v_y^2}$ = constant is a circle, so this division of the plane is into annular regions between circles whose successive radii are Δv apart:

Constant speed circles in a two-dimensional example

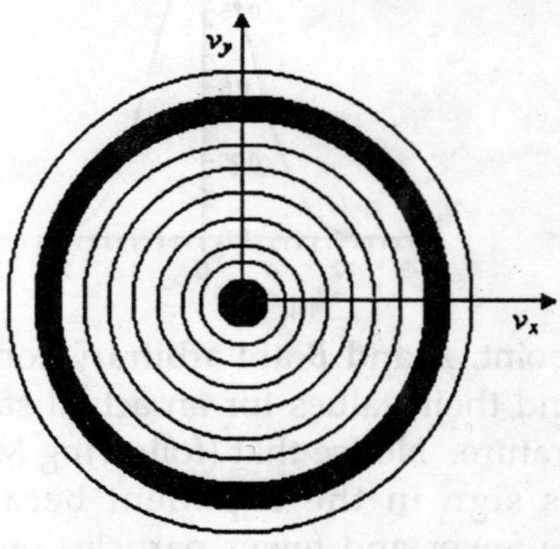

Constant speed circles in a two-dimensional example

Each of these annular areas corresponds to the same speed increment Δv. In particular, the green area, between a circle of radius 8Δv and one of radius 9Δv, corresponds to the same *speed increment* as the small red circle in the middle, which corresponds to speeds between 0 and Δv. Therefore, if the molecular speeds are pretty evenly distributed in this near-the-origin area of the (v_x, v_y) plane, there will be a lot more molecules with speeds between 8Δv and 9Δv than between 0 and Δv—so the *most likely speed* will *not* be zero. To find out what it actually is, we have to put this area argument together with the Gaussian fall off in density on going far from the origin. We'll discuss this shortly.

The same argument works in three dimensions—it's just a little more difficult to visualize. Instead of concentric circles, we have concentric spheres. All points lying on a spherical surface centered at the origin correspond to the same speed.

Let us now figure out the distribution of particles as a function of speed. The distribution in the three-dimensional space (v_x, v_y, v_z) is from Maxwell's analysis

# of particles in small box $dv_x dv_y dv_z$

$$= nf_1(v_x)f_1(v_y)f_1(v_z)dv_x dv_y dv_z$$

$$= NA^3 e^{-B\left(v_x^2+v_y^2+v_z^2\right)} dv_x dv_y dv_z$$

$$= NA^3 e^{-Br^2}\, dv_x dv_y dv_z$$

To translate this to the number of particles having *speed* between v and $v + dv$ we need to figure out how many of those little $dv_x dv_y dv_z$ boxes there are corresponding to speeds between v and $v + dv$. In other words, what is the *volume* of velocity space between the two neighbouring spheres, both centered at the origin, the inner one with radius v, the outer one infinitesimally bigger, with radius $v + dv$? Since dv is so tiny, this volume is just the area of the sphere multiplied by dv: that is, $4\pi v^2\, dv$.

Finally, then, the probability distribution as a function of speed is:

$$f(v)dv = 4\pi v^2\, A^3 e^{-Br2} dv.$$

Of course, our job isn't over—we still have these two unknown constants A and B. However, just as for the function $f_1(v_x)$, $f(v)dv$ is the fraction of the molecules corresponding to speeds between v and $v + dv$, and all these fractions taken together must add up to 1.

That is,

$$\int_0^\infty f(v)\,dv = 1$$

We need the standard result $\int_0^\infty x^2 e^{-Br^2} dx = (1/4B)\sqrt{\pi/B}$

(a derivation can be found in my 152 Notes on Exponential Integrals), and find:

$$4\pi A^3 \frac{1}{4B}\sqrt{\frac{\pi}{B}} = 1$$

This means that there is really only one arbitrary variable left: if we can find B, this equation gives us A: that is, $4\pi A^3 = \frac{4}{\sqrt{\pi}} B^{3/2}$, and $4\pi A^3$ is what appears in $f(v)$.

Looking at $f(v)$, we notice that B is a measure of how far the distribution spreads from the origin: if B is small, the distribution drops off more slowly—the average particle is more energetic. Recall now that the average kinetic energy of the particles is related to the temperature by $\overline{\frac{1}{2}mv^2} = \frac{3}{2}kT$. This means that B is related to the inverse temperature.

In fact, since $f(v)dv$ is the fraction of particles in the interval dv at v, and those particles have kinetic energy ½mv^2, we can use the probability distribution to find the average kinetic energy per particle:

$$\overline{\frac{1}{2}mv^2} = \int_0^\infty \frac{1}{2}mv^2 f(v)\,dv.$$

To do this integral we need another standard result:

$$\int_0^\infty x^4 e^{-Bx^2} dx = \left(3/8B^2\right)\sqrt{\pi/B}.$$

We find:

$$\overline{\frac{1}{2}mv^2} = \frac{3m}{4B}.$$

.Substituting the value for the average kinetic energy in terms of the temperature of the gas,

$$\overline{\frac{1}{2}mv^2} = \frac{3}{2}kT.$$

gives $B = m/2kT$, so $4\pi A^3 = \frac{4}{\sqrt{\pi}} B^{3/2} = 4\pi\left(\frac{m}{2\pi kT}\right)^{3/2}$.

This means the distribution function

$$f(v) = 4\pi\left(\frac{m}{2\pi kT}\right)^{\frac{3}{2}} v^2 e^{-mv^2/2kT} = 4\pi\left(\frac{m}{2\pi kT}\right)^{\frac{3}{2}} v^2 e^{-E/kT}$$

where E is the kinetic energy of the molecule.

Note that this function increases parabolically from zero for low speeds, reaches a maximum, then decreases exponentially. As the temperature increases, the position of the maximum shifts to the right. The total area under the curve is always one, by definition. For air molecules (say, nitrogen) at room temperature the curve is the blue one below. The red one is for an absolute temperature down by a factor of two:

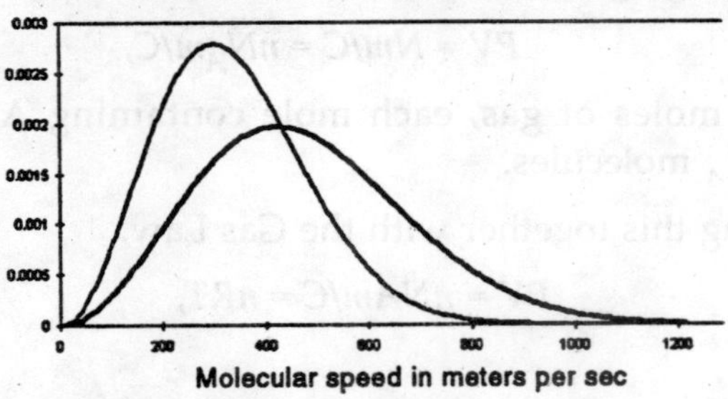

WHAT ABOUT POTENTIAL ENERGY?

Maxwell's analysis solves the problem of finding the statistical velocity distribution of molecules of an ideal gas in a box at a definite temperature T: the relative probability of a molecule having velocity $\vec{v}$ is proportional to $e^{-mv^2/2kT} = e^{-E/kT}$. The *position* distribution is taken to be uniform: the molecules are assumed to be equally likely to be anywhere in the box.

But how is this distribution affected if in fact there is some kind of potential pulling the molecules to one end of the box?

In fact, we've already solved this problem, in the discussion earlier on the isothermal atmosphere. Consider a really big box, kilometers high, so air will be significantly denser towards the bottom. Assume the temperature is uniform throughout. We found under these conditions that with Boyles Law expressed in the form

$$\rho = CP$$

the atmospheric density varied with height as

$P = P_0e^{-cgh}$, or equivalently $\rho = \rho_0e^{-cgh}$.

Now we know that Boyle's Law is just the fixed temperature version of the Gas Law Pv = nRT, and the density

$$\rho = \text{mass/volume} = Nm/V$$

with N the total number of molecules and m the molecular mass,

$$CP = \rho = Nm/V.$$

Rearranging,

$$PV = Nm/C = nN_Am/C,$$

for n moles of gas, each mole containing Avogadro's number N_A molecules.

Putting this together with the Gas Law,

$$PV = nNAm/C = nRT,$$

so

$$C = N_Am/RT = m/kT$$

where Boltzmann's constant $k = R/N_A$ as discussed previously.

The dependence of gas density on height can therefore be written

$$\rho = \rho_0e^{-Cgh} = \rho_0e^{-mgh/kT}.$$

The important point here is that mgh is the potential energy of the molecule, and the distribution we have found is exactly parallel to Maxwell's velocity distribution, the potential energy now playing the role that kinetic energy played in that case.

We're now ready to put together Maxwell's velocity distribution with this height distribution, to find out how the molecules are distributed in the atmosphere, *both* in velocity space *and* in ordinary space. In other words, in a *six*-dimensional space!

Our result is:

$$f(x,y,z,v_x,v_y,v_z) = f(h,v) \alpha e^{-mv2/2kT} e^{-mgh/kT}$$
$$= e^{-(1/2)mv2 + mgh)/kT} = e^{-E/kT}.$$

That is, *the probability of a molecule having total energy E is proportional to* $e^{-E/kT}$.

This is the Boltzmann, or Maxwell-Boltzmann, distribution. It turns out to be correct for any type of potential energy, including that arising from forces between the molecules themselves.

DEGREES OF FREEDOM AND EQUIPARTITION OF ENERGY

By a "degree of freedom" we mean a way in which a molecule is free to move, and thus have energy—in this case, just the x, y, and z directions. Boltzmann reformulated Maxwell's analysis in terms of degrees of freedom, stating that there was an average energy $\frac{1}{2}kT$ in each degree of freedom, to give total average kinetic energy $3.\frac{1}{2}kT$, so the specific heat per molecule is presumable $1.5k$, and given that $k = R/N_A$, the specific heat per mole comes out at $1.5R$. In fact, this is experimentally confirmed for monatomic gases. However, it is found that diatomic gases can have specific heats of $2.5R$ and even $3.5R$. This is not difficult to understand—these molecules have more degrees of freedom. A dumbbell molecule can rotate about two directions perpendicular to its axis. A diatomic molecule could also vibrate. Such a simple harmonic oscillator motion has both kinetic and potential energy, and it turns out to have total energy kT in thermal equilibrium. Thus, reasonable explanations for the specific heats of various gases can be concocted by assuming a contribution ½k from each degree of freedom. But there are problems. Why shouldn't the dumbbell rotate about its axis? Why do monatomic atoms not rotate at all? Even more ominously, the specific heat of hydrogen, 2.5R at room temperature, drops to 1.5R at lower temperatures. These problems were not resolved until the advent of quantum mechanics.

BROWNIAN MOTION

One of the most convincing demonstrations that gases really *are* made up of fast moving molecules is Brownian motion, the observed constant jiggling around of tiny particles, such as fragments of ash in smoke. This motion was first noticed by a Scottish botanist, who initially assumed he was looking at living creatures, but then found the same motion in what he knew to be particles of inorganic material. Einstein showed how to use Brownian motion to estimate the size of atoms.

Chapter 5

Ideal Gas Thermodynamics

IDEAL GAS THERMODYNAMICS

The Kinetic Theory picture of a gas (outlined in the previous lecture) is often called the *Ideal Gas Model*. It ignores interactions between molecules, and the finite size of molecules. In fact, though, these only become important when the gas is very close to the temperature at which it become liquid, or under extremely high pressure. In this lecture, we will be analyzing the behaviour of gases in the pressure and temperature range corresponding to heat engines, and in this range the Ideal Gas Model is an excellent approximation. Essentially, our programme here is to learn how gases absorb heat and turn it into work, and *vice versa*. This heat-work interplay is called *thermodynamics*.

Julius Robert Mayer was the first to appreciate that there *is* an equivalence between heat and mechanical work. The tortuous path that led him to this conclusion is described in an earlier lecture, but once he was there, he realised that in fact the numerical equivalence—how many Joules in one calorie in present day terminology—could be figured out easily from the results of some measurements of gas specific heat by French scientists. The key was that they had measured specific heats both at constant *volume* and at constant *pressure*. Mayer realised that in the latter case, heating the gas necessarily increased its volume, and the gas therefore did work in pushing to expand its container. Having convinced himself that mechanical work and heat were equivalent, evidently the extra heat needed to raise the temperature of the gas at constant pressure was exactly the work the gas did on its container. (*Historical note*: although he did the work in 1842, he didn't publish until 1845, and at first miscalculated—but then gave a figure within 1per cent of the correct value of 4.2 joules per calorie.)

The simplest way to see what's going on is to imagine the gas in a cylinder, held in by a piston, carrying a fixed weight, able to move up and down the cylinder smoothly with negligible friction. The pressure on the gas is just the total weight pressing down divided by the area of the piston, and this total weight, of course, will not change as the piston moves slowly up or down: the gas is at *constant pressure.*

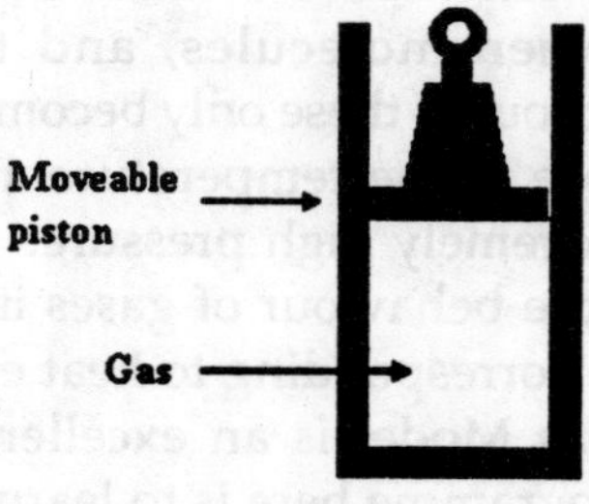

THE GAS SPECIFIC HEATS C_V AND C_P

Consider now the two specific heats of this same sample of gas, let's say one mole:

Specific heat at constant volume, C_V (piston glued in place),

Specific heat at constant pressure, C_P (piston free to rise, no friction).

In fact, we already worked out C_V in the Kinetic Theory lecture: at temperature *T*, recall the average kinetic energy per molecule is $\frac{3}{2}kT$, so one mole of gas—Avogadro's number of molecules—will have total kinetic energy, which we'll label *internal energy*,

$$E_{int} = \frac{3}{2}kT \cdot N_A = \frac{3}{2}RT.$$

(In this simplest case, we are ignoring the possibility of the molecules having their *own* internal energy: they might be spinning or vibrating—we'll include that shortly).

That the internal energy is $\frac{3}{2}RT$ per mole immediately gives us the specific heat of a mole of gas in a fixed volume,

$$C_V = \frac{3}{2}R$$

that being the heat which must be supplied to raise the temperature by one degree.

However, if the gas, instead of being in a fixed box, is held in a cylinder at constant pressure, experiment confirms that *more* heat must be supplied to raise the gas temperature by one degree. As Mayer realised, the total heat energy that must be supplied to raise the temperature of the gas one degree *at constant pressure* is $\frac{3}{2}k$ per molecule *plus* the energy required to lift the weight. The work the gas must do to raise the weight is the force the gas exerts on the piston multiplied by the distance the piston moves. If the area of piston is A, then the gas at pressure P exerts force PA. If on heating through one degree the piston rises a distance Δh, the gas does work

$$PA.\,\Delta h = P\Delta V.$$

Now, for one mole of gas, $PV = RT$, so at constant P

$$P\Delta V = P\Delta T.$$

Therefore, the work done by the gas in raising the weight is just $R\Delta T$, the specific heat at constant pressure, the total heat energy needed to raise the temperature of one mole by one degree,

$$C_P = C_V + R.$$

In fact, this relationship is true whether or not the molecules have rotational or vibrational internal energy. (It's known as Mayer's relationship.) For example, the specific heat of oxygen at constant volume

$$C_V(O_2) = \frac{5}{2}R$$

and this is understood as a contribution of $\frac{3}{2}R$ from kinetic energy, and R from the two rotational modes of a dumbbell molecule (just why there is no contribution form rotation about the third axis can only be understood using quantum mechanics). The specific heat of oxygen at constant pressure

$$G_P(O_2) = \frac{1}{2}R.$$

It's worth having a standard symbol for the *ratio* of the specific heats:

$$\frac{C_P}{C_V} = \gamma.$$

ISOTHERMS AND ADIABATS

An ideal gas in a box has three thermodynamic variables: P, V, T. But if there is a fixed mass of gas, fixing two of these variables fixes the third from $PV = nRT$ (for n moles). In a heat engine, heat can enter the gas, then leave at a different stage. The gas can expand doing work, or contract as work is done on it. To track what's going on as a gas engine transfers heat to work, say, we must follow the varying state of the gas. We do that by tracing a curve in the (P, V) plane. Supplying heat to a gas which consequently expands and does mechanical work is the key to the heat engine. But just knowing that a gas is expanding and doing work is not enough information to follow its path in the (P, V) plane. The route it follows will depend on whether or not heat is being supplied (or taken away) at the same time. There are, however, two particular ways a gas can expand reversibly—meaning that a tiny change in the external conditions would be sufficient for the gas to retrace its path in the (P, V) plane backwards. It's important to concentrate on reversible paths, because as Carnot proved and we shall discuss later, they correspond to the most efficient engines. The two sets of reversible paths are the *isotherms* and the *adiabats*.

Isothermal behaviour: The gas is kept *at constant temperature* by allowing heat flow back and forth with a very large object (a "heat reservoir") at temperature T. From $PV = nRT$, it is evident that for a fixed mass of gas, held at constant T but subject to (slowly) varying pressure, the variables P, V will trace a hyperbolic path in the (P, V) plane.

This path, $PV = nRT_1$, say is called the *isotherm* at temperature T_1. Here are two examples of isotherms:

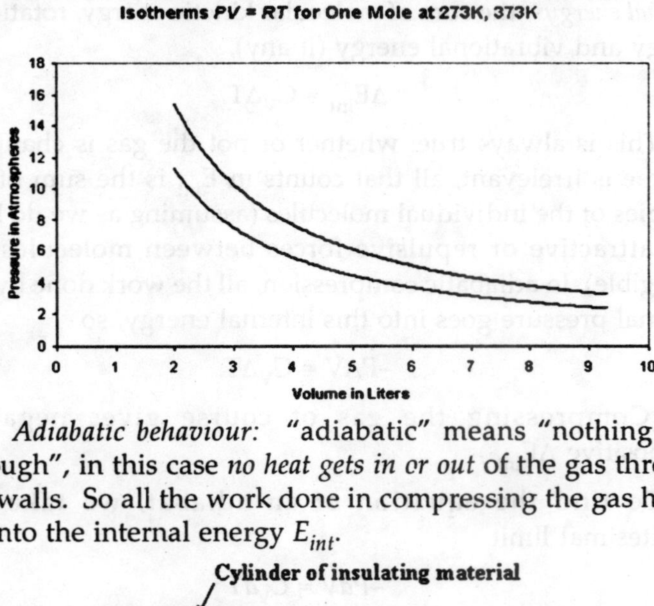

Adiabatic behaviour: "adiabatic" means "nothing gets through", in this case *no heat gets in or out* of the gas through the walls. So all the work done in compressing the gas has to go into the internal energy E_{int}.

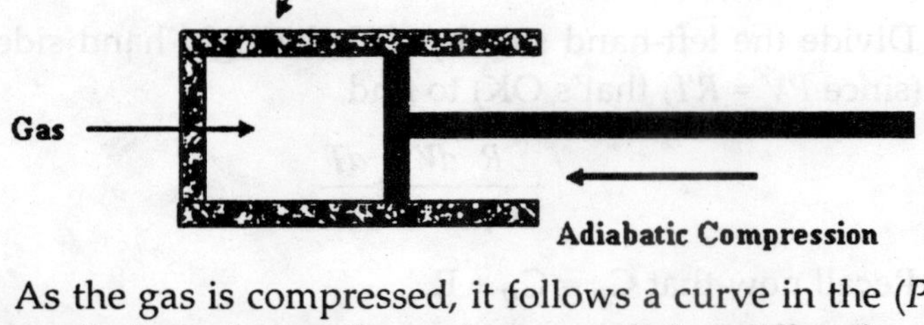

As the gas is compressed, it follows a curve in the (P, V) plane called an *adiabat*. To see how an adiabat differs from an isotherm, imagine beginning at some point on the blue 273K isotherm on the above graph, and applying pressure so the gas moves to higher pressure and lower volume. Since the gas's internal energy is increasing, but the number of molecules is staying the same, its temperature is necessarily rising, it will move towards the red curve, then above it. This means *the adiabats are always steeper than the isotherms.*

EQUATION FOR AN ADIABAT

What equation for an adiabat corresponds to $PV = nRT_1$ for an isotherm?

On raising the gas temperature by ΔT, the change in the *internal energy*—the sum of molecular kinetic energy, rotational energy and vibrational energy (if any),

$$\Delta E_{int} = C_V \Delta T.$$

This is always true: whether or not the gas is changing volume is irrelevant, all that counts in E_{int} is the sum of the energies of the individual molecules (assuming as we do here that attractive or repulsive forces between molecules are negligible). In adiabatic compression, all the work done by the external pressure goes into this internal energy, so

$$-P\Delta V = C_V \Delta T.$$

(Compressing the gas of course gives negative ΔV, positive ΔE_{int}.)

To find the equation of an adiabat, we take the infinitesimal limit

$$-PdV = C_V dT$$

Divide the left-hand side by PV, the right-hand side by RT (since $PV = RT$, that's OK) to find

$$-\frac{R}{C_V}\frac{dV}{V} = \frac{dT}{T}.$$

Recall now that $C_P = C_V + R$,

and $C_P/C_v = \gamma$. It follows that

$$\frac{R}{C_V} = \frac{C_P - C_V}{C_V} = \gamma - 1.$$

Hence

$$-(\gamma - 1) \int \frac{dv}{v} = \int \frac{dT}{T}$$

and integrating

$\ln T + (\gamma - 1) \ln V = const.$

from which the equation of an adiabat is

TV

$TV^{\gamma-1}$ = const.

From $PV = RT$, the P, V equation for an adiabat can be found by multiplying the left-hand side of this equation by the constant PV/T, giving

PV^{γ} = const. for an adiabat,

where $\gamma = 5/3$ for a monatomic gas, $7/5$ for a diatomic gas.

Chapter 6

Conduction, Convection, Radiation

In the simplest of terms, the discipline of heat transfer is concerned with only two things: temperature, and the flow of heat. Temperature represents the amount of thermal energy available, whereas heat flow represents the movement of thermal energy from place to place.

On a microscopic scale, thermal energy is related to the kinetic energy of molecules. The greater a material's temperature, the greater the thermal agitation of its constituent molecules (manifested both in linear motion and vibrational modes). It is natural for regions containing greater molecular kinetic energy to pass this energy to regions with less kinetic energy.

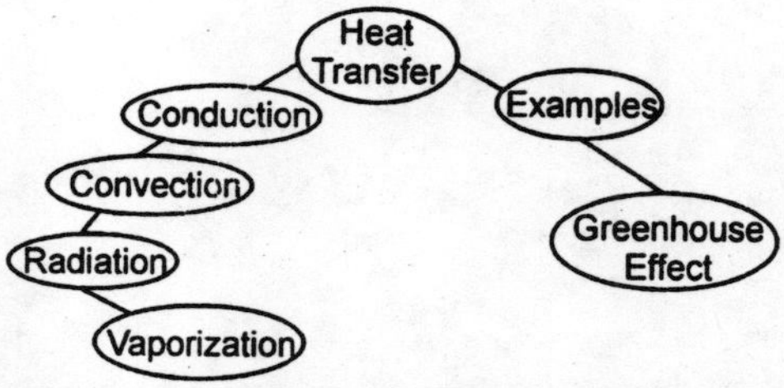

Several material properties serve to modulate the heat transferred between two regions at differing temperatures. Examples include thermal conductivities, specific heats, material densities, fluid velocities, fluid viscosities, surface emissivities, and more. Taken together, these properties serve to make the solution of many heat transfer problems an involved process.

The transfer of heat is normally from a high temperature object to a lower temperature object. Heat transfer changes the internal energy of both systems involved according to the First Law of Thermodynamics.

HEAT CONDUCTION

Conduction is heat transfer by means of molecular agitation within a material without any motion of the material as a whole. If one end of a metal rod is at a higher temperature, then energy will be transferred down the rod toward the colder end because the higher speed particles will collide with the slower ones with a net transfer of energy to the slower ones. For heat transfer between two plane surfaces, such as heat loss through the wall of a house, the rate of conduction heat transfer is:

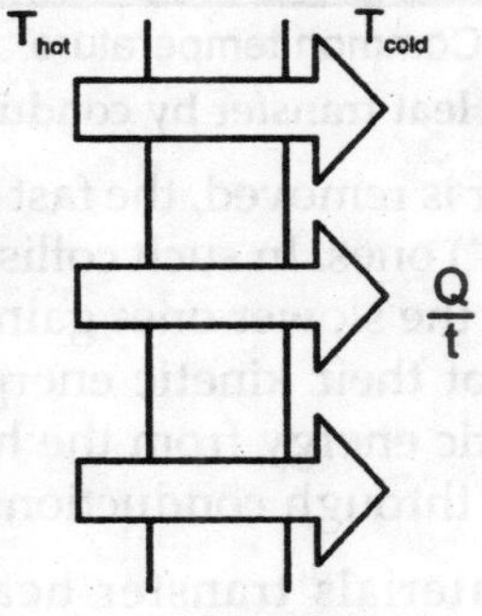

$$\frac{Q}{t} = \frac{kA(T_{hot} - T_{cold})}{d}$$

Q = heat transferred in time = t

k = thermal conductivity of the barrier

A = area

T = temperature

d = thickness of barrier

The flow of heat by conduction occurs via collisions between atoms and molecules in the substance and the subsequent transfer of kinetic energy. Let us consider two substances at different temperatures separated by a barrier which is subsequently removed, as in the following figure.

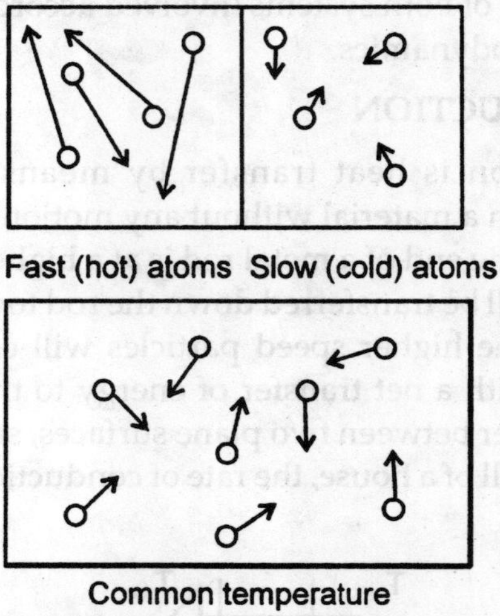

Fig. Heat transfer by conduction

When the barrier is removed, the fast ("hot") atoms collide with the slow ("cold") ones. In such collisions the faster atoms lose some speed and the slower ones gain speed; thus, the fast ones transfer some of their kinetic energy to the slow ones. This transfer of kinetic energy from the hot to the cold side is called a flow of heat through conduction.

Different materials transfer heat by conduction at different rates - this is measured by the material's thermal conductivity. Suppose we place a material in between two reservoirs at different temperatures, as in the following figure.

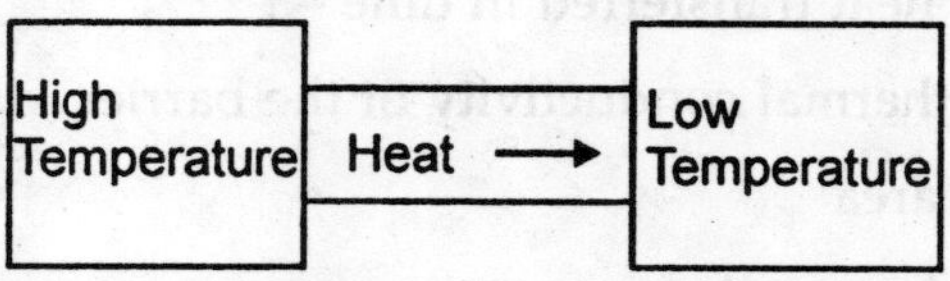

Fig. Measurement of thermal conductivity

Let us now measure the flow of heat through the material over time. Knowing the material's cross-sectional area and length, the thermal conductivity of the material is then defined as.

Heat flow over time = Thermal conductivity × (area/ length) × (temperature difference between reservoirs)

Thus, for a given temperature difference between the reservoirs, materials with a large thermal conductivity will transfer large amounts of heat over time - such materials, like copper, are good thermal conductors. Conversely, materials with low thermal conductivities will transfer small amounts of heat over time - these materials, like concrete, are poor thermal conductors. This is why if you throw a piece of copper and a piece of concrete into a campfire, the copper will heat up much more quickly than the concrete. It is also why fiberglass insulation, and also feathers and fur, have air pockets - dead air is a poor thermal conductor, and so the air pockets aid in cutting back on the heat loss through the material.

Home insulation is thus a poor thermal conductor, which keeps as much heat in as possible. Instead of being rated in terms of thermal conductivity, insulation is therefore usually rated in terms of its thermal resistance, which is defined as

Thermal resistance = length/thermal conductivity

Materials which have a high thermal conductivity have, by definition, a low thermal resistance - they are poor heat insulators. On the other hand, materials with a low thermal conductivity have a high thermal resistance - they are good heat insulators. Good insulating materials therefore should have a high thermal resistance. In fact, the "R" value quoted for insulation is the thermal resistance (in British units).

HEAT CONVECTION

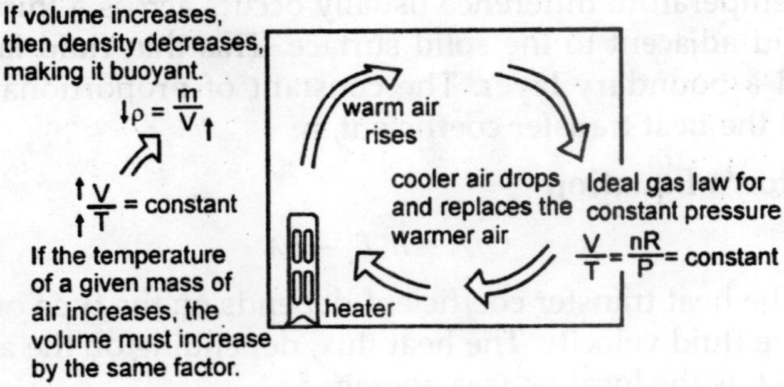

Convection is heat transfer by mass motion of a fluid such as air or water when the heated fluid is caused to move away from the source of heat, carrying energy with it. Convection above a hot surface occurs because hot air expands, becomes less dense, and rises. Hot water is likewise less dense than cold water and rises, causing convection currents which transport energy.

Convection can also lead to circulation in a liquid, as in the heating of a pot of water over a flame. Heated water expands and becomes more buoyant. Cooler, more dense water near the surface descends and patterns of circulation can be formed, though they will not be as regular as suggested in the drawing.

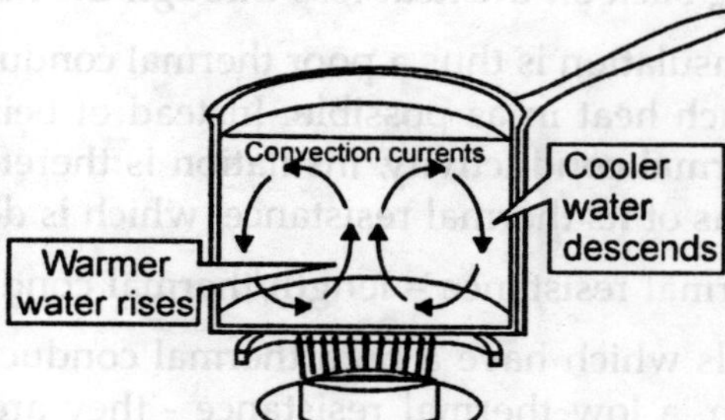

Convection heat transfer is energy transport due to bulk fluid motion. Convection heat transfer through gases and liquids from a solid boundary results from the fluid motion along the surface.

Newton determined that the heat transfer/area, Q/A, is proportional to the fluid solid temperature difference Ts-Tf. The temperature difference usually occurs across a thin layer of fluid adjacent to the solid surface. This thin fluid layer is called a boundary layer. The constant of proportionality is called the heat transfer coefficient, h.

Newton's Equation

$$Q/A = h(T_s - T_f)$$

The heat transfer coefficient depends on the type of fluid and the fluid velocity. The heat flux, depending on the area of interest, is the local or area averaged.

RADIATION

Radiation heat transfer is energy transport due to emission of electromagnetic waves or photons from a surface or volume. The radiation does not require a heat transfer medium, and can occur in a vacuum. Radiation is the only form of heat transfer that can occur in the absence of any form of medium and as such is the only means of heat transfer through a vacuum. Thermal radiation is a direct result of the movements of atoms and molecules in a material. Since these atoms and molecules are composed of charged particles (protons and electrons), their movements result in the emission of electromagnetic radiation, which carries energy away from the surface. At the same time, the surface is constantly bombarded by radiation from the surroundings, resulting in the transfer of energy to the surface. Since the amount of emitted radiation increases with increasing temperature, a net transfer of energy from higher temperatures to lower temperatures results.

The frequencies of the emitted photons are described by the Plank distribution. A black body at higher temperature will emit photons having a distributional peak at a higher frequency than will a colder object, and their respective spectral peaks will be separated according to Wien's displacement law. The photosphere of the Sun, at a temperature of approximately 6000 K, emits radiation principally in the visible portion of the spectrum. The solar radiation incident upon the earth's atmosphere is largely passed through to the surface. The atmosphere is largely transparent in the visible spectrum. However, in the infrared spectrum that is characteristic of a blackbody at 300K, the temperature of the earth, the atmosphere is largely opaque. The blackbody radiation from earth's surface is absorbed or scattered by the atmosphere. Though some radiation escapes into space, it is the radiation absorbed and subsequently emitted by atmospheric gases. It is this spectral selectivity of the atmosphere that is responsible for the planetary greenhouse effect.

The behaviour of a common household lightbulb has a spectrum overlapping the blackbody spectra of the sun and the earth. A portion of the photons emitted by a tungsten light bulb filament at 3000K lie in the visible spectrum. However, the majority of the photonic energy is associated with longer wavelengths and will transfer heat to the environment, as can be deduced empirically by observing a household incandescent lightbulb. Whenever EM radiation is emitted and then absorbed, heat is transferred. This principle is used in microwave ovens, laser cutting, and RF hair removal.

The heat transfer by radiation is proportional to the fourth power of the absolute material temperature. The proportionality constant σ is the Stefan-Boltzman constant equal to 5.67×10^{-8} W/m²K⁴. The radiation heat transfer also depends on the material properties represented by ε, the emissivity of the material.

$$Q/A = \varepsilon\sigma T^A$$

For a surface with an emissivity of ε = 0.8 and T = 373 K (100C), the radiation heat transfer is

$$Q/A = \varepsilon\sigma T^A = (0.8)\ (5.67 \times 10^{-8})\ (373)^4 = 878\ W/m^2$$

For moderate (less than 100 C) temperature differences, it should be noted that the radiation and natural convection heat transfer are about the same.

SPECIFIC HEAT

The specific heat (also called specific heat capacity) is the amount of heat required to change a unit mass (or unit quantity, such as mole) of a substance by one degree in temperature. Therefore, unlike the extensive variable heat capacity, which depends on the quantity of material, specific heat is an intensive variable and has units of energy per mass per degree (or energy per number of moles per degree).

The heat capacity of a substance can differ depending on what extensive variables are held constant, with the quantity being held constant usually being denoted with a subscript. For example, the specific heat at constant pressure is commonly denoted C_p, while the specific heat at constant volume is commonly denoted C_V. The specific heat of water at constant atmospheric pressure is

$$C_P = 1 \text{ cal g}^{-1} \text{ K}^{-1} = 4.814 \text{ J g}^{-1} \text{ K}^{-1}$$

i.e., 1 calorie is needed per degree Kelvin (or Celsius) of temperature change for 1 gram of liquid water. In fact, the definition of (one of the several types of) the calorie is the amount of heat needed to change the temperature of 1 g of water by 1°C its temperature of maximum density (roughly 3.98° C).

The specific heat capacity of a solid or liquid is defined as the heat required to raise a unit mass of substance by one degree of temperature. This can be stated by the following equation:

$$\Delta Q = m\, c\, \Delta T$$

where,

ΔQ = Heat supplied to substance,

m = Mass of the substance,

c = Specific heat capacity,

ΔT = Temperature rise.

Suppose that a body absorbs an amount of heat ΔQ and its temperature consequently rises by ΔT. The usual definition of the heat capacity, or specific heat, of the body is

$$c = \frac{\Delta Q}{\Delta T} \quad (1)$$

If the body consists of ν moles of some substance then the molar specific heat (i.e., the specific heat of one mole of this substance) is defined

$$c = \frac{1}{\nu}\frac{\Delta Q}{\Delta T} \quad (2)$$

In writing the above expressions, we have tacitly assumed that the specific heat of a body is independent of its temperature. In general, this is not true. We can overcome this problem by only allowing the body in question to absorb a very small amount of heat, so that its temperature only rises slightly, and its specific heat remains approximately constant.

In the limit as the amount of absorbed heat becomes infinitesimal, we obtain

$$c = \frac{1}{\nu}\frac{đQ}{dT}$$

In classical thermodynamics, it is usual to define two specific heats. Firstly, the molar specific heat at constant volume, denoted

$$c = \frac{1}{\nu}\left(\frac{đQ}{dT}\right)_V,$$

and, secondly, the molar specific heat at constant pressure, denoted

$$c_P = \frac{1}{\nu}\left(\frac{đQ}{dT}\right)_P.$$

Consider the molar specific heat at constant volume of an ideal gas. Since $dV = 0$, no work is done by the gas, and the first law of thermodynamics reduces to

$$đQ = dE.$$

It follows from Eq. that

$$c_V = \frac{1}{\nu}\left(\frac{\partial E}{\partial T}\right)_V$$

Now, for an ideal gas the internal energy is volume independent. Thus, the above expression implies that the specific heat at constant volume is also volume independent. Since E is a function only of T, we can write

$$dE = \left(\frac{\partial E}{\partial T}\right)_V dT$$

The previous two expressions can be combined to give

$$dE = \nu\, c_V\, dT$$

for an ideal gas.

Let us now consider the molar specific heat at constant pressure of an ideal gas. In general, if the pressure is kept constant then the volume changes, and so the gas does work on its environment. According to the first law of thermodynamics,

$$đQ = dE + p\,dV = \nu\, c_V\, dT + p\,dV.$$

The equation of state of an ideal gas tells us that if the volume changes by dV, the temperature changes by dT, and the pressure remains constant, then

$$p\,dV = \nu R\,dT$$

The previous two equations can be combined to give

$$đQ = \nu\ c_V\, dT + \nu R\,dT$$

Now, by definition

$$c_P = \frac{1}{\nu}\left(\frac{đQ}{dT}\right)_P,$$

so we obtain

$$c_P = c_V + R$$

for an ideal gas. Note that at constant volume all of the heat absorbed by the gas goes into increasing its internal energy, and, hence, its temperature, whereas at constant pressure some of the absorbed heat is used to do work on the environment as the volume increases. This means that, in the latter case, less heat is available to increase the temperature of the gas. Thus, we expect the specific heat at constant pressure to exceed that at constant volume, as indicated by the above formula.

The ratio of the two specific heats C_p/C_V is conventionally denoted γ. We have

$$\gamma = \frac{c_P}{c_V} = 1 + \frac{R}{c_V}$$

for an ideal gas. In fact, γ is very easy to measure because the speed of sound in an ideal gas is written

$$c_s = \sqrt{\frac{\gamma p}{\rho}},$$

where ρ is the density.

Table: Specific heat capacities of some common substances

Substance	*Phase*	$c_p J\ g^{-1} K^{-1}$
Air (Sea level, dry, 0 °C)	gas	1.0035
Aluminium	solid	0.897
Ammonia	liquid	4.700
Argon	gas	0.5203
Beryllium	solid	1.82
Copper	solid	0.385
Diamond	solid	0.5091
Ethanol	liquid	2.44
Gold	solid	0.1291
Graphite	solid	0.710
Helium	gas	5.1932
Hydrogen	gas	14.30
Iron	solid	0.450
Lithium	solid	3.58
Mercury	liquid	0.1395
Nitrogen	gas	1.040
Neon	gas	1.0301
Oxygen	gas	0.918
Silica (fused)	solid	0.703
Uranium	solid	0.116
Water	gas (373 K)	1.8369
liquid (298 K)	4.1813	
solid (273 K)	2.0870	

LAB EXPERIMENT TO DETERMINE SPECIFIC HEAT CAPACITY OF SUBSTANCES

The material is in the form of metal particles or "shot." This is heated to steam temperature, then dumped into cool water in the calorimeter and allowed to come to an equilibrium temperature. The necessary measurements are: mass of water, m_w; initial water temperature, Tw; mass of metal sample, m_s; sample temperature, Ts; and the final equilibrium temperature of the mixture, T_f. The heat exchange equation is:

$$m_w(T_w - T_f) + m_s s_s(T_s - T_f) + m_c s_c(T_c - T_f) = 0$$

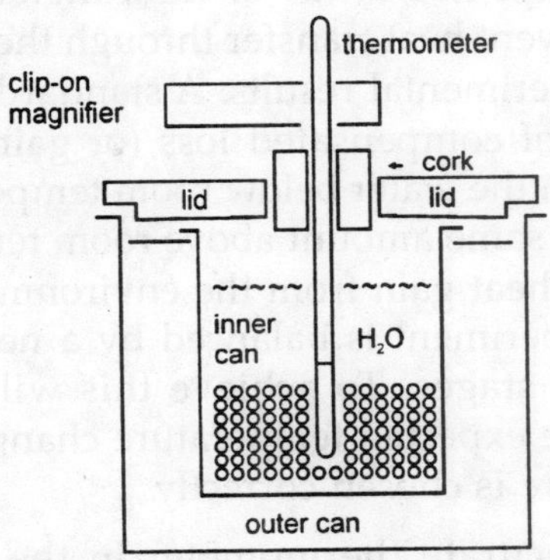

Fig. Double-walled calorimeter

The subscript c labels quantities which describe the calorimeter inner cup. This cup also participates in the heat exchange. Its temperature is nearly the same as the water in it, so $T_c = T_w$.

Several kinds of calorimeter may be available:

1. A styrofoam cup with a styrofoam lid. This plastic material contains many cells of trapped air and is a remarkably effective insulator.
2. The double-walled calorimeter, which consists of two metal cans, usually aluminium, separated by an insulating fibre ring at the top, and covered with a wooden or plastic lid. The trapped air between the cans provides insulation. This space may also be filled with fibreglass.
3. The vacuum calorimeter is a Dewar vacuum flask (similar to a Thermos (TM) bottle). The inside wall is silvered to prevent radiative transfer to the outside; the vacuum prevents conduction.

4. The "adiabatic wall" container in which the walls are kept at the same temperature as the contents inside. This is accomplished by temperature sensors, electrical heaters, and an electrical feedback system which senses any slight temperature difference and quickly corrects it.

When the first two kinds of calorimeter are used, it is important to prevent heat transfer through the container from affecting the experimental results. A standard way to do this is the "method of compensated loss (or gain)." The idea is simple: start with the water below room temperature and end with it about the same amount above room temperature. This ensures that the heat gain from the environment in the early stages of the experiment is balanced by a nearly equal heat loss in the later stages. To achieve this will require some knowledge of the expected temperature change, so the initial water temperature is chosen correctly.

Insert an accurate thermometer in the calorimeter to monitor the temperature of the water in the inner cup. Take care that the thermometer is inserted to the proper depth. Most thermometers have a line engraved around the stem, indicating the exact depth to which the thermometer should be immersed in liquid to insure accurate temperature values.

Some calorimeters have a stirring plunger. While temperature readings are being made the water should be gently and continually stirred. Move the plunger up and down with slow and regular strokes. Maintain uniform stirring both before and after adding the shot. Avoid stirring too vigorously, for that can add significant mechanical energy to the system!

One still must consider the heat transfer to the inner cup and the thermometer. Both are in contact with the water and can exchange heat with it. The "compensated loss" procedure only corrected for heat exchange with the outer environment, but did nothing to correct for heat exchanges within the calorimeter.

We know the materials of the cup and thermometer (aluminium and glass). We can weigh the cup, and can estimate the fraction of the glass of the thermometer which participates in heat exchange with the water. Then, using handbook values of the specific heat capacities of aluminium and glass we can calculate the size of the terms which contribute in the heat exchange equations.

$$m_c s_c (T_w - T_f) + m_t s_t (T_w - T_f)$$

Since these bodies follow the water temperature, they share the (Tw – Tf) factor with the water term of the heat transfer equation. We can therefore combine all three terms:

$$(m_2 + m_c s_c + m_{ct} s_t)(T_2 - T_f)$$

The quantity $(m_s s_c + m_t s_t) = m_{eq}$ is often called the "water equivalent" of the cup and thermometer. Once it is determined, this number is simply added to the measured mass of water, for the purposes of calculation of the heat transfer equation:

$$(m_w + m_{eq})(T_2 - T_f) + m_s s_s (T_s - T_f) = 0$$

The water equivalent correction will be small, and need not be of high accuracy unless the temperature differences can be measured with high accuracy. The error analysis will show how accurate this must be.

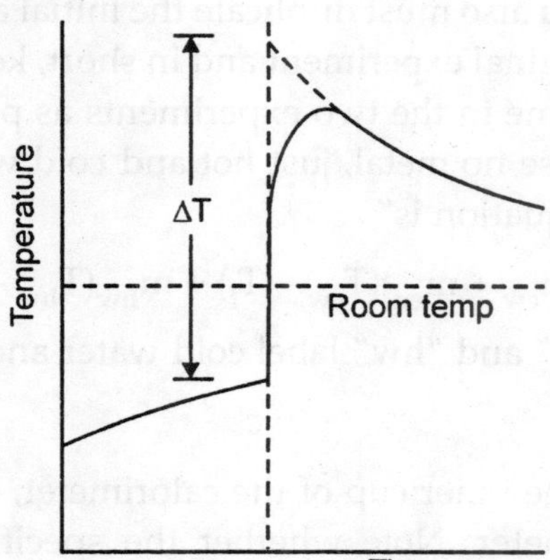

Fig. Method of temperature correction

The final experimental difficulty to be overcome is the thermal "sluggishness" of the thermometer. If you watched carefully as the hot metal was added to the water, you noticed that the thermometer reading did not change immediately. It took time for the thermal energy to transfer to the thermometer fluid. By the time the thermometer reached equilibrium with the water, the water had already cooled somewhat by energy loss to the surroundings. This "thermal lag" error of the thermometer might cause the calculated final temperature to be too low.

One can correct for thermometer sluggishness in the following way. Monitor the water temperature at half-minute intervals for about 5 minutes before the hot metal is added, and also for about 5 minutes after it is added. When this temperature data is graphed against time, The "Newtonian" cooling and warming curves are evident, and can be extrapolated back to the time the shot was dumped in.

If the temperature measurements are very accurate, you may need to determine the water equivalent by a better method. One way is to perform a separate experiment in which hot water is added to cold water in the calorimeter in such amounts as to duplicate the temperature change of the original experiment. You also must duplicate the initial and final water levels of the original experiment and in short, keep everything as much the same in the two experiments as possible, except this time you use no metal, just hot and cold water. Now the heat-transfer equation is"

$$(m_{cw} + m_{eq})(T_{cw} - T_f) + m_{hw}(T_{hw} - T_f) = 0$$

where "cw" and "hw" label cold water and hot water.

Method

1. Weigh the inner cup of the calorimeter, and weigh the thermometer. Note whether the specific heat values of the calorimeter cup is stamped on it. If it is not, determine the metal of which it is made, and estimate its specific heat value.
2. Determine what mass of water would be required to fill the inner cup 2/3 full.

3. Do a trial calculation to determine what mass of shot at 100° C must be added to cold water at 5° below room temperature to give a final equilibrium temperature about 5° above room temperature. Since you don't yet know the specific heat of the metal, you will have to make an intelligent guess of its value for this calculation. Consult handbook values for metals, but keep in mind the possibility that the sample you have been given just might be an alloy whose value is not in the tables. We want to achieve a temperature change of at least 10° in water, without using so much shot that the cup becomes overfull.
4. Now that you have determined how much shot to use, weigh out that much. Put it in the steam-generator cup, plug the top of the cup with cotton, and heat the shot in the steam- generator.
5. Put the cold water in the calorimeter inner cup and weigh both together. Assemble the calorimeter and begin monitoring the water temperature every 1/2 minute. Use the limited range (0–50° C) thermometer here, with the clip-on magnifier, for accurate readings.
6. When the shot cup has been in the steam bath for at least 5 minutes the metal shot will be at the steam temperature (near 100° C depending on the barometric pressure). Quickly transfer the shot into the calorimeter and reassemble it, then continue your temperature readings every of the water in the calorimeter every 1/2 minute. The exact time of dumping in the shot must be noted also. Take temperature readings for long enough to clearly establish the cooling curve of the water.
7. At the conclusion of the experiment, put the wet shot in the designated place to be dried. Do not put wet shot back in the original container. Empty the water from the boiler and the calorimeter. Dry any spills on the table and floor.

Analysis

1. Solve the heat-exchange equation algebraically for the specific heat of the unknown sample. Include the calorimeter cup's contribution to the heat exchange.
2. Compare the errors in the specific heat determination which would be caused by (a) a 0.1° error in measuring the temperature change of the water, and (b) a 0.1° error in determining the temperature change of the metal sample.
3. Suppose you had neglected to include the term representing the calorimeter inner cup. What percent change would this have made in the calculated value of the specific heat of the specimen? Would the specific heat value have been too high, or too low? How would this systematic error compare with the other experimental errors?
4. If all other sources of error were negligible, how accurate would the temperature measurements have to be to ensure 2per cent accuracy in the determination of the specific heat of the metal?

Examples:

1. Calculate the amount of heat needed to increase the temperature of 250g of water from 20°C to 56°C.

$$q = m \times C_g \times (T_f - T_w)$$

$$m = 250g$$

$$C_g = 4.18 \text{ J } ^\circ C^{-1} g^{-1}$$

$$T_f = 56^\circ C$$

$$T_w = 20^\circ C$$

$$q = 250 \times 4.18 \times (56 - 20)$$

$$q = 250 \times 4.18 \times 36$$

$$q = 37\,620 \text{ J} = 38 \text{ kJ}$$

2. Calculate the specific heat capacity of copper given that 204.75 J of energy raises the temperature of 15g of copper from 25° to 60°.

$$q = m \times C_g \times (T_f - T_w)$$
$$q = 204.75\ J$$
$$m = 15g$$
$$T_w = 25^oC$$
$$T_f = 60^oC$$
$$204.75 = 15 \times C_g \times (60 - 25)$$
$$204.75 = 15 \times C_g \times 35$$
$$204.75 = 525 \times C_g$$
$$C_g = 204.75 \div 204.75 = 0.39\ J^oC^{-1}\ g^{-1}$$

3. 216 J of energy is required to raise the temperature of aluminium from 15° to 35°C. Calculate the mass of aluminium.(Specific Heat Capacity of aluminium is 0.90 $J^oC^{-1}g^{-1}$).

$$q = m \times C_g \times (T_f - T_w)$$
$$q = 216\ J$$
$$C_g = 0.90\ J^oC^{-1}g^{-1}$$
$$T_w = 15^oC$$
$$T_f = 35^oC$$
$$216 = m \times 0.90 \times (35 - 15)$$
$$216 = m \times 0.90 \times 20$$
$$216 = m \times 18$$
$$m = 216 \div 18 = 12g$$

4. The initial temperature of 150g of ethanol was 22°C. What will be the final temperature of the ethanol if 3240 J was needed to raise the temperature of the ethanol? (Specific heat capacity of ethanol is 2.44 $J^oC^{-1}g^{-1}$).

$$q = m \times C_g \times (T_f - T_w)$$
$$q = 3240\ J$$
$$m = 150g$$

$$C_g = 2.44\ \text{J}^\circ\text{C}^{-1}\text{g}^{-1}$$

$$T_w = 22^\circ\text{C}$$

$$3240 = 150 \times 2.44 \times (T_f - 22)$$

$$3240 = 366\ (T_f - 22)$$

$$8.85 = T_f - 22$$

$$T_f = 30.9^\circ\text{C}$$

FOURIER'S LAW

Heat conduction is the transmission of heat across matter.

Heat transfer is always directed from a higher to a lower temperature. Denser substances are usually better conductors; metals are excellent conductors.

The law of heat conduction, also known as Fourier's law, states that the time rate of heat flow Q through a slab (or a portion of a perfectly insulated wire, as shown in the figure) is proportional to the gradient of temperature difference:

$$\frac{\Delta Q}{\Delta t} \sim kA\frac{\Delta T}{\Delta x}$$

Q is the time rate of heat flow through a slab,

Δt is the change in time,

k is a conductivity constant (dependent on the nature of the material and its temperature),

A is the transversal surface area,

ΔT is the temperature difference through which the heat is being transferred,

Δx is the thickness of the body of matter through which the heat is passing.

This law forms the basis for the derivation of the heat equation. R-value is the unit for heat resistance, the reciprocal of the conductance.

Writing

$$U = \frac{k}{\Delta x},$$

Fourier's law can also be stated as:

$$Q = U\,A\,\Delta T$$

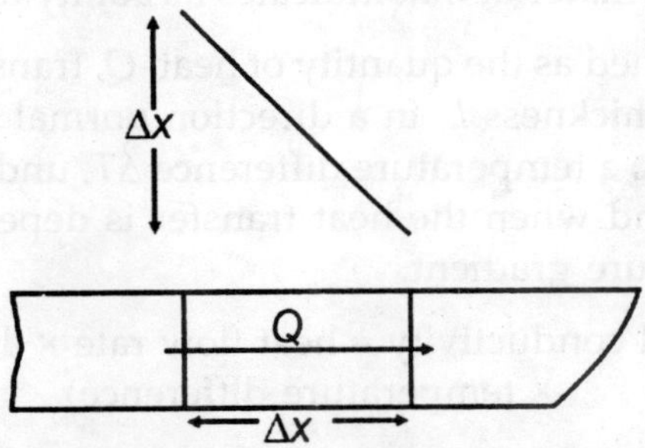

where U is the conductance. The reciprocal of conductance is resistance, equal to:

$$\frac{A\Delta T}{Q},$$

and it is resistance which is additive when several conducting layers lie between the hot and cool regions, because A and Q are the same for all layers. In a multilayer partition, the total conductance is related to the conductance of its layers by:

$$\frac{1}{U} = \frac{1}{U_1} + \frac{1}{U_2} + \frac{1}{U_3} + \ldots$$

So, when dealing with a multilayer partition, the following formula is usually used:

$$Q = \frac{A\Delta T}{\frac{\Delta_1 x}{K_1} + \frac{\Delta_2 x}{K_2} + \frac{\Delta_3 x}{K_3} \ldots}$$

When heat is being conducted from one fluid to another through a barrier, it is sometimes important to consider the conductance of the thin film of fluid which remains stationary next to the barrier. This thin film of fluid is difficult to quantify, its characteristics depending upon complex conditions of turbulence and viscosity, but when dealing with thin high-conductance barriers it can sometimes be quite significant.

Thermal Conductivity

In physics, thermal conductivity, k, is the intensive property of a material that indicates its ability to conduct heat.

It is defined as the quantity of heat, Q, transmitted in time t through a thickness L, in a direction normal to a surface of area A, due to a temperature difference ΔT, under steady state conditions and when the heat transfer is dependent only on the temperature gradient.

Thermal conductivity = heat flow rate × distance / (area × temperature difference)

$$k = \frac{Q}{t} \times \frac{L}{A \times \Delta T}$$

The units of K are $J\ s^{-1}\ m^{-1}\ K^{-1}$ or $W\ m^{-1}\ K^{-1}$.

Example values of the coefficient of thermal conductivity

Silver	428 $W\ m^{-1}\ K^{-1}$	Metals are good conductors, due to
Aluminium	236 $W\ m^{-1}K^{-1}$	their free electrons.
Steel	~50 $W\ m^{-1}\ K^{-1}$	
Glass	~1.0 $W\ m^{-1}\ K^{-1}$	Gases, and solids which trap gases,
are		
Wood	0.15 $W\ m^{-1}\ K^{-1}$	poor conductors of heat.
Air	0.024 $W\ m^{-1}\ K^{-1}$	

In metals, thermal conductivity approximately tracks electrical conductivity, as the freely moving valence electrons transfer not only electric current but also heat. However, this correlation is not applicable to all materials, highly electrically conductive silver is shown to be less thermally conductive than diamond, which is an electrical semiconductor. The reason for this difference is that phonons, not electrons, are the primary carriers of heat in diamond.

Thermal conductivity is not a simple property, and depends intimately on structure and temperature. For instance, pure crystalline substances also exhibit highly variable thermal conductivities along different crystal axes, due to differences in phonon coupling along a given crystal dimension. Air and other gases are generally good insulators, in the absence of convection. Therefore, many insulating materials function simply by having a large number of gas-filled pockets which prevent large-scale convection.

Thermal conductivity is clearly an important quantity for construction and related fields. However, materials used in such trades are rarely subjected to chemical purity standards.

Problems

1. The air in a room has a temperature of 28°C and the air outside is at 0°C. What is the rate of heat loss per unit area through a glass window pane of thickness 4 mm, the thermal conductivity of the glass being 0.80 J m^{-1} s^{-1} K^{-1}?
2. The glass of the window of a room is 5 mm thick and has an area of 4 m^2. If heat can only escape through the window, calculate the electric power required to keep the room at a uniform temperature 20°C above the external temperature. Repeat the calculation for a double glazed window consisting of two 5 mm thick glass sheets separated by non-convecting air 2.5 mm thick.

   [Thermal conductivity of glass is 1.5 W m^{-1} K^{-1}, of air 21 × 10^{-3} W m^{-1} K^{-1}]
3. A copper block of mass 200 g is suspended in a vacuum by a copper wire 10 cm long and 1 mm in diameter. If the copper is initially 100°C above its surroundings, find the rate at which it begins to cool, assuming that radiation can be neglected.

   [Specific heat of copper is 380 J kg^{-1} K^{-1}, Thermal con-ductivity of copper is 400 W m^{-1} K^{-1}.]
4. A lake in winter is covered by a layer of ice 20 cm thick. At night, the air temperature above the lake is –15°C. Calculate the rate at which the ice thickens in

mm per hour. (Assume there is no heat transfer between the ground below the lake and the water.)

[Thermal conductivity of ice is 2.2$Wm^{-1}K^{-1}$, Latent heat of ice is 336x10^3 J kg^{-1}, Density of ice is 920 kg m^{-3}.]

EXPERIMENT TO VERIFY FOURIER'S LAW FOR CONDUCTION IN SOLID HOMOGENOUS MATERIALS

Conduction Along a Simple Bar

Conduction is defined as the transfer of energy from more energetic particles to adjacent less energetic particles as a result of interactions between the particles. In solids, conduction is the combined result of molecular vibrations and free electron mobility. Metals typically have high free electron mobility, which explains why they are good heat conductors.

Conduction can be easily understood if we imagine two blocks, one very hot and the other cold. If we put these blocks in contact with one another and insulate them, we will observe energy in the form of heat being transferred from the hot block to the cold block. This heat transfer between the two solid blocks is conduction.

Theory

In this experiment we will investigate conduction in an insulated long slender brass bar. We will assume that the bar is of length L, a uniform hot temperature This imposed on one end, and a cold temperature Tc is imposed on the other. We will also assume, because the bar is insulated, that all the heat flows in the axial direction due to an imposed temperature differential along the bar.

The thermal conductivity k varies between different materials and can be a function of temperature, but it can be treated as a constant over small temperature ranges. Because of the enhancement of heat transfer by free electrons, thermal conductivity is analogous to electrical conductivity and as a result, metals that are good conductors of electricity are also good conductors of heat.

In this experiment we will investigate Fourier's Law by finding the thermal conductivity k for brass and comparing this value to the actual value.

Apparatus

The apparatus we will be using in this experiment is the P.A. Hilton H940 Heat Conduction Unit, which consists of three items. The first item is a transformer equipped with a circuit breaker. The transformer has two cords which connect it to an AC outlet and to the second item, the calibration unit.

The calibration unit has two basic functions. It delivers power to the heater element within the test unit and it calibrates and displays the temperatures at nine locations along the test unit. The amount of power delivered to the test unit is controlled by the power control knob on the right side of the calibration unit; to its left is the temperature selector knob which is used to select one of the nine thermocouple temperatures for display on the digital readout.

The third item is the test unit, which consists of two test geometries: an insulated brass bar which allows a sample to be placed between the two ends, and an insulated disk. Both geometries are equipped with a power supply, but we will only use the insulated bar for this experiment. The test unit is also equipped with a cooling water hose. The purpose of the cooling water running through the unit at the cold end of the bar is to remove heat that is produced at the hot end and transferred by conduction to the cold end, keeping the cold end at roughly a constant temperature. Once the rate at which heat is generated is equal to the rate at which heat is removed, steady state equilibrium conditions will exist. At this point the temperatures will be constant throughout the bar and readings can be taken. The test unit has two heater cords: one from the test bar and the other from the test disk. Be sure to use only the heater cord for the bar, which connects to the heater plug located in the lower right hand corner of the calibration unit.

In addition to this equipment, you will also need a box that contains the samples, thermocouples, and conducting paste used in the procedure. The samples can be placed in the bar test unit by releasing the clamps and sliding the cold end of the bar out. The samples will fit into the test unit in two different ways, but only one will ensure good contact between the sample and the test unit. The conducting paste is a highly conductive compound which is designed to decrease contact resistance when applied to the ends of the connecting bars.

Method

1. Connect the equipment, making sure that the calibration unit is switched off before connecting the transformer to the AC outlet.
2. Liberally coat each side of a brass sample of 30 mm length and 25 mm diameter and the connecting surfaces of the test unit with conducting compound.
3. Insert the test sample into the unit and allow cooling water to flow through the test unit.
4. Switch the calibration unit on and adjust the power control knob to deliver 5 W of power to the test unit; allow the system to reach steady state (approximately 20 minutes).
5. Record the temperatures at each of the nine thermocouples and the power input.
6. Repeat steps 4 and 5 for power settings of 10 W and 15 W, each time allowing approximately 20 minutes for the system to reach steady state.
7. Plot the temperature profile along the length of the bar for each trial and determine the slopes of the best-fit lines.
8. Calculate the thermal conductivity k of the brass and steel for each trial. Compute the average value of k and compare it to the reference value for cartridge brass.

9. Discuss the characteristics of your plots and compare them to what you would expect based on the theory. Is Fourier's Law satisfied? Also discuss the validity of the assumptions made and sources of error within the equipment or through measurements, etc.

Chapter 7

Steam

Vapour is the gaseous state of matter. Although vapour and gas are frequently used interchangeably, vapour often carries the connotation of gaseous matter in a state of equilibrium with identical matter in a liquid or solid state below its boiling point. A liquid or solid does not have to boil to release a vapour. The constituent atoms or molecules of a vapour possess vibrational, rotational, and translational motion. Vapour phase is an intermediate stage between the solid and the perfect gas state. Important fluids in this group are H_2O (i.e. Steam) and the refrigerants.

Steam

Steam refers to vapourized water. It is a pure, completely invisible gas. Pure steam, (unmixed with air, but in equilibrium with water-liquid), has a temperature of around 100 degrees Celsius at standard atmospheric pressure, and occupies about 1,600 times the volume of liquid water (steam can of course be much hotter than the boiling point of water; such steam is usually called superheated steam). In the atmosphere, the partial pressure of water is much lower than 1 atm, therefore gaseous water can exist at temperatures much lower than 100 C.

In common speech, steam most often refers to the white mist that condenses above boiling water as the hot vapour ("steam" in the first sense) mixes with the cooler air. This mist is made of tiny droplets of liquid water, not gaseous water, so it is no longer technically steam.

Refrigerant

A refrigerant is a compound used in a heat cycle that undergoes a phase change from a gas to a liquid and back. The two main uses of refrigerants are refrigerators/freezers and air conditioners. The ideal refrigerant has good thermodynamic properties, is noncorrosive, and safe. The desired thermodynamic properties are a boiling point somewhat below the target temperature, a high heat of vapourization, a moderate density in liquid form, and a relatively high density in gaseous form. Since boiling point and gas density are affected by pressure, refrigerants may be made more suitable for a particular application by choice of operating pressure.

Corrosion properties are a matter of materials compatibility with the components used for the compressor, piping, evaporator, and condenser. Safety considerations include toxicity and flammability. Early mechanical refrigeration systems employed sulphur dioxide gas or anhydrous ammonia, with small home refrigerators primarily using the former. Being toxic, sulphur dioxide rapidly disappeared from the market with the introduction of Freon. Ammonia is still used in some large commercial plants, well away from residential areas, where a leak will not cause widespread injuries.

Until concerns about depletion of the ozone layer arose in the 1980s, the most widely used refrigerants were the halomethanes R-12 and R-22, with R-12 being more common in automotive air conditioning and small refrigerators, and R-22 being used for residential and light commercial air conditioning, refrigerators, and freezers. Some very early systems used R-11 because its low boiling point allows low-pressure systems to be constructed, reducing the mechanical strength required for components. New production of R-12 ceased in the United States in 1995, and R-22 is to be phased out in 2010. R-134a and certain blends are now replacing chlorinated compounds.

VAPOUR PRESSURE

Vapour pressure is the pressure of the vapour over a liquid (and some solids) at equilibrium. The gas that is above the liquid is called its vapour and it creates a pressure called vapour pressure.

Each molecule in the liquid has energy, but not the same amount. The energy is distributed according to the Maxwell-Boltzmann distribution. All the molecules are in motion because of their energy, but none have sufficient energy to break the mutual attractive force molecules have for each other. Suppose, that surface molecule was moving up away from the surface and had enough energy to break away from the attractive forces of the molecules around it. Where would that molecule go? It would continue to move away from the liquid surface and it becomes a molecule of gas. It happens to another molecule and another and another. As more and more molecules leave the surface, some return to the surface and resume their former life as a liquid molecule. Soon the number of molecules in the vapour phase is constant because the rate of returning equals the rate of leaving and so the pressure stays constant.

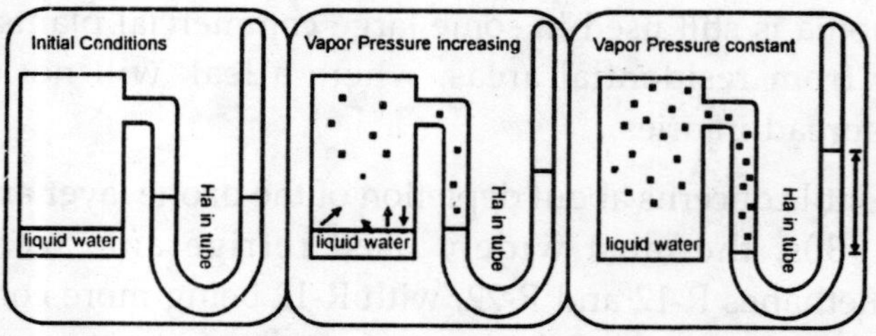

The vapour pressure depends only on the temperature. That is true because as the temperature goes up, there are more and more molecules with the right combination of energy and direction to break free of the liquid's surface. Finally, we might think that, if we were to put more liquid in, the vapour pressure would go up. That's not correct. Keep in mind that there are two opposing processes at work:

1. Molecules leaving the liquid's surface and
2. Molecules returning to the liquid at its surface.

Vapour Pressure of Water

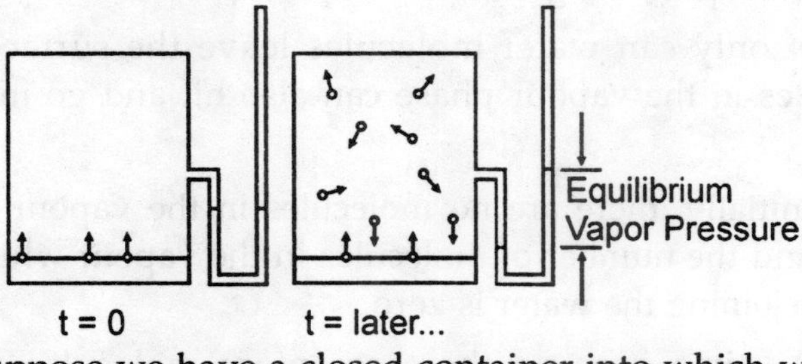

Suppose we have a closed container into which we pour some water. As soon as we add the water we check a pressure gauge connected to the container. We let the container sit for a while and then we check the pressure again. As the water evaporates the pressure exerted by the vapour above the liquid increases, until at some point, the pressure reaches a constant value, the vapour pressure of the substance:

The Molecular Basis

The kinetic energy of the molecules at the surface of a liquid varies over a range of values:

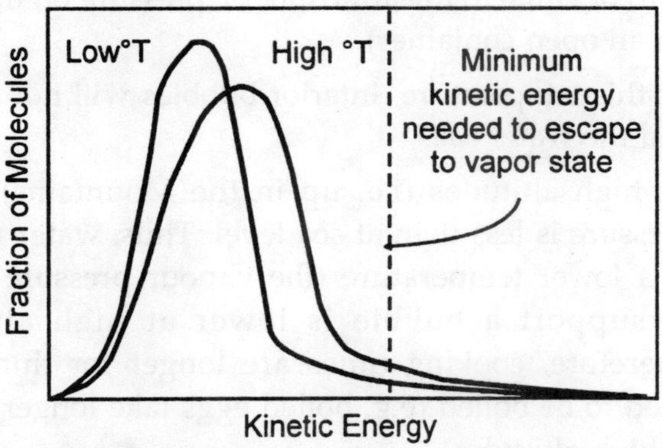

1. Some of the molecules have enough kinetic energy to overcome the attractive forces between the molecules
2. The weaker the attractive forces, the greater the fraction of molecules with enough kinetic energy to escape.

3. The greater the fraction of molecules which can escape the liquid, the greater the vapour pressure

Not only can water molecules leave the surface, but molecules in the vapour phase can also hit and go into the water

1. Initially, there are no molecules in the vapour phase and the number of molecules in the vapour which are rejoining the water is zero
2. As time goes on there are more molecules in the vapour phase and the number of a vapour molecule striking the water increases
3. At some point in time the number of vapour molecules rejoining the water equals the number leaving to go into the vapour phase. an equilibrium has been reached, and the pressure has stabilized at the characteristic vapour pressure of the substance

Vapour Pressure Increases with Increasing Temperature

1. At 100°C the vapour pressure of water is 760 torr (1 atm) or equal to the atmospheric pressure on the liquid (in an open container)
2. At this temperature, interior bubbles will not collapse and the water *boils*
3. At high altitudes (i.e. up in the Mountains) the air pressure is less than at sea level. Thus, water will boil at a lower temperature (the vapour pressure needed to support a bubble is lower at high altitude). Therefore, cooking times are longer for things that need to be boiled (e.g. boiled eggs take longer to cook at high altitudes).

A phase change may be written as a chemical reaction. The transition from liquid water to steam, for example, may be written as

$$\text{H2}\ (l) \rightleftarrows H_2\ (g)$$

The equilibrium constant for this reaction (the vapourization reaction) is

$$K = P_w$$

where P_w is the partial pressure of the water in the gas phase when the reaction is at equilibrium. This pressure is often called the *vapour pressure.* The vapour pressure is literally the partial pressure of the compound in the gas.

This equilibrium may be established at any temperature. Because vapourization reactions are endothermic, an increase in temperature will shift the equilibrium to the right. Thus at low temperatures the vapour pressure of the liquid is very low and at high temperatures the vapour pressure is quite large.

At What Temperature will the Liquid Boil?

The boiling point corresponds to the temperature at which the vapour pressure of the liquid equals the atmospheric pressure. If the liquid is open to the atmosphere (that is, not in a sealed vessel), it is not possible to sustain a pressure greater than the atmospheric pressure, because the vapour will simply expand until its pressure equals that of the atmosphere.

The temperature at which the vapour pressure exactly equals one atm is called the *normal boiling point.*

In a sealed vessel, where the vapour cannot expand and thus the pressure can build up, it is possible to establish the vapourization equilibrium at temperatures in excess of the normal boiling point.

The Van't Hoff equation provides a relationship between an equilibrium constant and temperature.

$$\ln K = -\frac{\Delta H_{vap}}{RT} + \frac{\Delta S_{vap}}{R}$$

For this reaction, the equilibrium constant is simply the vapour pressure, P, which when substituted into the above equation yields the Claussius-Clapeyron equation.

$$\ln P = -\frac{\Delta H_{vap}}{RT} + \frac{\Delta S_{vap}}{R}$$

The normal boiling point, $T_{bp}{}^{o}$, corresponds to the temperature at which both the reactant and the product are in the standard state. A pure liquid under 1 atm pressure is in the standard state. A pure gas at 1 atm pressure is also in the standard state. Thus in the standard state P = 1 atm. This relation allows the Claussius-Clapeyron equation to be rewritten as

$$\ln P = \frac{-\Delta H_{vap}}{\mathrm{R}}\left(\frac{1}{T} - \frac{1}{T_{bp}}\right)$$

and

$$T_{bp}{}^{o} = \frac{-\Delta H_{vap}}{\Delta S_{vap}}$$

As these equations illustrate, experiment data for the vapour pressure of a liquid as a function of temperature enables the calculation of both the normal boiling point (or indeed any boiling point), the standard enthalpy of vapourization, and the standard entropy of vapourization.

Experiment

- To determine the standard enthalpy of vapourization for ethanol.
- To determine the standard entropy of vapourization for ethanol.
- To determine the normal boiling point for ethanol.

The experimental equipment consists of a glass bulb connected to a manometer. The bulb is evacuated (that is, a vacuum pump is employed to remove all the gas inside the bulb) and a small amount of liquid ethanol is added. The bulb and manometer constitute a closed system containing liquid ethanol and gas-phase ethanol. Thus the pressure measured by the manometer is attributable solely to the enthanol vapour.

The temperature of the bulb and manometer can be varied. As the temperature changes, the vapour pressure of ethanol changes. The manometer may be used to measure the pressure at any temperature. The pressure-temperature data is automatically plotted on the graph at the right.

Record at least five data points (temperature and pressure) over the temperature range 0 to 50°C. A plot of ln *P* vs 1/*T* is created and the line-of-best-fit is plotted through the data. Use the slope and intercept of the line-of-best-fit to determine the standard enthalpy and entropy of vapourization and the normal boiling point of ethanol.

VAPOUR ENERGY

Why does boiling a liquid create vapour that can spin a turbine? Actually, all liquids have a certain "vapour pressure" at certain temperatures. For instance, a cup of water at room temperature has vapour pressure even though you don't see steam coming out of it. Let's go back to the cup of water example. Look at the surface of the water. Now imagine zooming in really close. If you could see the individual molecules, you would notice that they were all moving around slightly, constantly jostling and shoving each other but staying in the same overall position. The movement is caused by the energy associated with the given temperature.

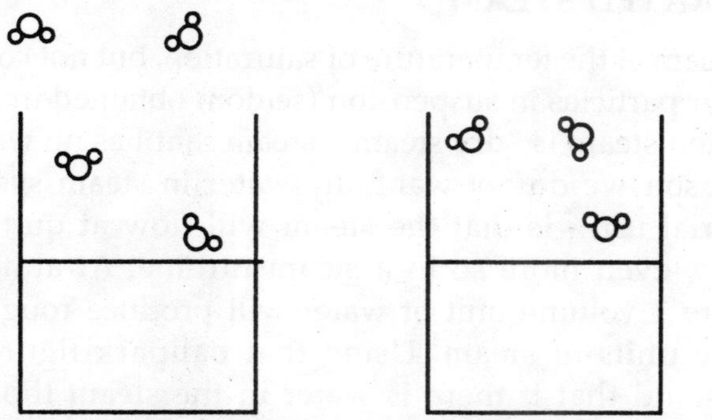

When there is no lid, water molecules can escape until there is eventually no water left in the container.

With a lid, the molecules are trapped in the container and so there is no net loss of water

At room temperature, the molecules will all have a certain amount of energy and will therefore be moving around at a certain speed, whereas at a colder temperature, the molecules will have less energy and will not be moving around as much. At absolute 0 (0 Kelvin, or –273 degrees Celsius), the lowest possible temperature, technically the movement of the molecules would be at a minimum (although there would still be movement).

Now imagine that one of the molecules of water near the surface is pushed a little harder from below. The molecule gains just enough energy to "break free" of the other molecules and is launched into the air. The free molecule will most likely escape into the open air if there is no lid on the container; if there is a lid, however, then the molecule will bounce off of the different surfaces within the system and eventually return to the surface of the liquid where it "reattaches" itself to the other molecules.

VAPOUR VOLUME

Vapour Volume is the number of cubic feet of vapour resulting from the complete evaporation of one gallon of liquid.

SATURATED STEAM

Steam at the temperature of saturation, but not containing of water particles in suspension (seldom obtained in general). Saturated steam is "dry steam", steam that has no water in it. The reason we do not want any water in steam system - in industrial use - is that the steam will flow at quite a high velocity, even more so in a steam turbine. At atmospheric pressure 1 volume unit of water will produce roughly 3600 volume units of steam. Using that ballpark figure, you'll understand that if there is water in the steam those water droplets would act like bullets (high density relative to the density of the steam, moving at speed), damaging equipment. Many turbines (turbines with a vacuum condensor) are designed in such a way that the "used" steam contains some water (up to about 10 per cent). The reason for this is that the condensation energy of that steam / water is used and thus gives a more energy efficient system.

Table: Thermodynamic Properties of Saturated Steam

0.0	14.696	212.00	180.07	970.3	1150.4	26.80
1.3	16.0	216.32	184.42	967.6	1152.0	24.75
2.3	17.0	219.44	187.56	965.5	1153.1	23.39
5.3	20.0	227.96	196.16	960.1	1156.3	20.09
10.3	25.0	240.07	208.42	952.1	1160.6	16.30
15.3	30.0	250.33	218.82	945.3	1164.1	13.75
20.3	35.0	259.28	227.91	939.2	1167.1	11.90
25.3	40.0	267.25	236.03	933.7	1169.7	10.50
30.3	45.0	274.44	243.36	928.6	1172.0	9.40
40.3	55.0	287.07	256.30	919.6	1175.9	7.79
50.3	65.0	297.97	267.50	911.6	1179.1	6.66
60.3	75.0	307.60	277.43	904.5	1181.9	5.82
70.3	85.0	316.25	286.39	897.8	1184.2	5.17
80.3	95.0	324.12	294.56	891.7	1186.2	4.65
90.3	105.0	331.36	302.10	886.0	1188.1	4.23
100.0	114.7	337.90	308.80	880.0	1188.8	3.88
110.3	125.0	344.33	315.68	875.4	1191.1	3.59
120.3	135.0	350.21	321.85	870.6	1192.4	3.33
125.3	140.0	353.02	324.82	868.2	1193.0	3.22
130.3	145.0	355.76	327.70	865.8	1193.5	3.11
140.3	155.0	360.50	333.24	861.3	1194.6	2.92
150.3	165.0	365.99	338.53	857.1	1195.6	2.75
160.3	175.0	370.75	343.57	852.8	1196.5	2.60
180.3	195.0	379.67	353.10	844.9	1198.0	2.34
200.3	215.0	387.89	361.91	837.4	1199.3	2.13
225.3	240.0	397.37	372.12	828.5	1200.6	1.92
250.3	265.0	406.11	381.60	820.1	1201.7	1.74
-	300.0	417.33	393.84	809.0	1202.8	1.54
-	400.0	444.59	424.00	780.5	1204.5	1.16
-	450.0	456.28	437.20	767.4	1204.6	1.03
-	500.0	467.01	449.40	755.0	1204.4	0.93
-	600.0	486.21	471.60	731.6	1203.2	0.77
-	900.0	531.98	526.60	668.8	1195.4	0.50
-	1200.0	567.22	571.70	611.7	1183.4	0.36
-	1500.0	596.23	611.60	556.3	1167.9	0.28
-	1700.0	613.15	636.30	519.6	1155.9	0.24
-	2000.0	635.82	671.70	463.4	1135.1	0.19
-	2500.0	668.13	730.60	360.5	1091.1	0.13

SUPER HEATED STEAM

Super heated steam is steam that has a higher temperature than saturated steam of the same pressure; the main reasons to use superheated steam are that it prevents condensation (which could otherwise occur with even a small energy loss) and that the steam carries more energy that way.

Thermodynamic Properties of Super heated steam

Absolute pressure (kN/m^2)	Saturation temperature (°C)	Steam Temperature (°C)						
		120	150	180	200	230	250	280
150	111.4	7,239	7,419	7,557	7,644	7,767	7,845	7,957
200	120.2		7,279	7,420	7,507	7,631	7,710	7,822
250	127.4		7,169	7,311	7,400	7,525	7,604	7,717
350	138.9		6,998	7,146	7,237	7,364	7,444	7,558
400	143.6		6,929	7,079	7,171	7,299	7,380	7,495
500	151.8			6,965	7,059	7,190	7,272	7,388
600	158.8			6,869	6,966	7,100	7,183	7,300
700	165.0			6,786	6,886	7,022	7,107	7,225
800	170.4			6,712	6,815	6,954	7,040	7,156
900	175.4			6,645	6,751	6,893	6,980	7,101
1,000	179.9			6,584	6,692	6,838	6,926	7,049
1,100	184.4				6,638	6,787	6,876	7,001
1,200	188.0				6,587	6,739	6,831	6,956
1,400	195.0				6,494	6,653	6,748	6,877
1,600	201.4					6,577	6,674	6,806
2,000	212.4					6,440	6,546	6,685
2,500	223.9					6,292	6,407	6,558
3,500	242.5						6,173	6,349

WET STEAM

Mixture of liquid and steam of the same substance in which both are at saturation temperature. If additional heat is added to the wet steam at constant pressure, the temperature remains constant until all liquid is evaporated (saturated steam); it is only at this point that the temperature increases above the saturation temperature (superheated steam).

Dryness fraction of Wet Steam

If the water content of the steam is 5per cent by mass, then the steam is said to be 95per cent dry and has a dryness fraction of 0.95.

Dryness fraction can be expressed as:

$$x = w_s / (w_w + w_s)$$

where

x = dryness fraction

w_w = mass of water (kg, lb)

w_s = mass of steam (kg, lb)

Enthalpy of Wet Steam

The actual enthalpy of evaporation of wet steam is the product of the dryness fraction – *x* - and the specific enthalpy – h_s - from the steam tables. Wet steam have lower usable heat energy than dry saturated steam.

$$h_t = h_s x + h_w$$

where

h_t= enthalpy of wet steam (kJ/kg, Btu/lb)

h_s = enthalpy of steam (kJ/kg, Btu/lb)

h_w = enthalpy of saturated water or condensate (kJ/ kg, Btu/lb)

Specific Volume of Wet Steam

The droplets of water in wet steam will occupy negligible space in the steam and the specific volume of wet steam will be reduced according the dryness fraction.

$$v = v_s x$$

where

v = specific volume of wet steam (m^3/kg, ft^3/lb)

v_s = specific volume of the dry steam (m^3/kg, ft^3/ lb)

DRY STEAM

Steam which does not contain water held in suspension mechanically; — sometimes applied to superheated steam. A measure of the "wetness" of steam is called quality. Quality varies from 0 to 1, with a quality of 0 being a saturated liquid and 1 being a saturated vapour. When steam has a quality of 1, it is considered "dry" steam. If the quality falls between 0 and 1, it is considered "wet" steam. You can have a can of steam because of the ideal gas law, PV = nRT. At a given pressure (P), volume (V), and temperature (T), a certain mass (n is the number of moles of a substance) can be in just about any state. Tip the pressure one way, and the state changes. Same goes for volume and temperature. The trick to get a dry can of steam is to know the volume of your can and the amount of H_20 in the can. Assuming the can to be a constant volume and that the H_2O cannot escape, you pick the temperature and pressure needed to get "dry" steam.

FLASH STEAM

The term 'flash steam' is traditionally used to describe steam issuing from condensate receiver vents and open-ended condensate discharge lines from steam traps. How can steam be formed from water without adding heat?

Flash steam occurs whenever water at high pressure (and a temperature higher than the saturation temperature of the low-pressure liquid) is allowed to drop to a lower pressure. Conversely, if the temperature of the high-pressure water is lower than the saturation temperature at the lower pressure, flash steam cannot be formed. In the case of condensate passing through a steam trap, it is usually the case that the upstream temperature is high enough to form flash steam.

Fig. Flash steam formed because T1 > T2

Consider a kilogram of condensate at 5 bar g and a saturation temperature of 159°C passing through a steam trap to a lower pressure of 0 bar g. The amount of energy in one kilogram of condensate at saturation temperature at 5 bar g is 671 kJ. In accordance with the first law of thermodynamics, the amount of energy contained in the fluid on the low-pressure side of the steam trap must equal that on the high-pressure side, and constitutes the principle of conservation of energy.

Consequently, the heat contained in one kilogram of low-pressure fluid is also 671 kJ. However, water at 0 bar g is only able to contain 419 kJ of heat, subsequently there appears to be an imbalance of heat on the low-pressure side of 671 – 419 = 252 kJ, which, in terms of the water, could be considered as excess heat.

This excess heat boils some of the condensate into what is known as flash steam and the boiling process is called flashing. Therefore, the one kilogram of condensate which existed as one kilogram of liquid water on the high pressure side of the steam trap now partly exists as both water and steam on the low-pressure side.

The amount of flash steam produced at the final pressure (P_2) can be determined using given Equation

$$\text{Proportion of flash steam} = \frac{(h_f \text{ at } P_1) - (h_f \text{ at } P_2)}{h_{fg} \text{ at } P_2}$$

Where:

P_1 = Initial pressure

P_2 = Final pressure

h_f = Liquid enthalpy (kJ/kg)

h_{fg} = Enthalpy of evaporation (kJ/kg)

Example The case where the high pressure condensate temperature is higher than the low pressure saturation temperature.

Consider a quantity of water at a pressure of 5 bar g, containing 671 kJ/kg of heat energy at its saturation temperature of 159°C. If the pressure was then reduced down to atmospheric pressure (0 bar g), the water could only exist at 100°C and contain 419 kJ/kg of heat energy. This difference of 671 - 419 = 252 kJ/kg of heat energy, would then produce flash steam at atmospheric pressure.

Therefore:

$$\text{Flash steam produced} = \frac{670.9 - 419.0}{2257.0}$$

Total flash steam = 0.11kg steam/kg water

The proportion of flash steam produced can be thought of as the ratio of the excess energy to the enthalpy of evaporation at the final pressure.

DRYNESS FRACTION

The dryness fraction of the steam is defined as the ratio Weight of dry steam/Weight of stuff. If a pure liquid is heated at a constant pressure there is a fixed temperature at which bubbles of vapour form in the liquid; this phenomenon is known as boiling. The states of substance at this condition represents a point on the properties diagram, known as boiling point; e.g. point 1. A slight addition of heat to the liquid at this state changes some of it into vapour.

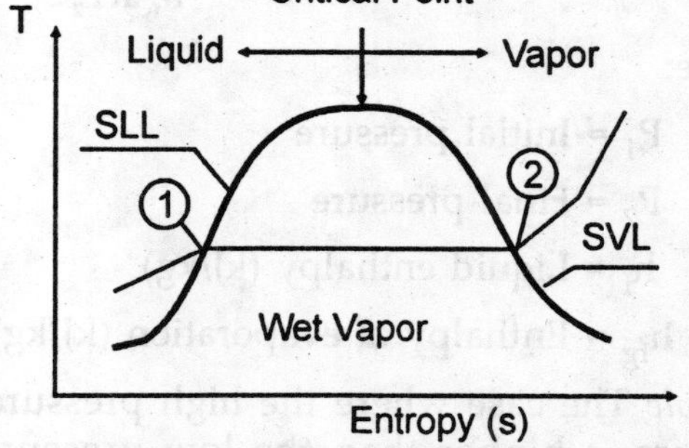

Saturated liquid line, SLL, is formed by connecting a series of boiling points. Boiling temperature known also as saturation temperature, T, for a pure liquid is only a function of pressure, P, i.e.

$$T_2 = T_1 = f(P)$$

Vapourization continues by further heat supply to the system until no liquid is left. This state is known as dry saturated vapour, e.g. point 2. If the system is slightly cooled at this state, then droplets of liquid will begin to form. Connecting a series of points at dry saturated vapour builds a line, known as saturated vapour line, SVL. The state of substance between saturated liquid and dry vapour is called wet vapour. Further heating of a dry saturated vapour at constant pressure causes a rise of vapour temperature and it becomes superheated. The state of substance is completely defined by its pressure and temperature if it is in liquid or superheated vapour phase i.e.

$$h = f1(P,T) = \text{Specific enthalpy}$$

$$v = f2(P,T) = \text{Specific volume}$$

$$s = f3(p,T) = \text{Specific entropy}$$

These properties for different substances are either tabulated or can be calculated by certain equations, e.g. IFC formulation for water and steam properties. The state of wet vapour can not be defined by just pressure and temperature until one other property is given. The condition or quality of wet vapour is often defined by its dryness or wetness fraction.

Dryness fraction,

x = The mass of dry vapour in 1 kg of the mixture,

and,

Wetness fraction,

$1 - x$ = The mass of liquid in 1 kg of the mixture.

For wet vapour with the dryness fraction, x,

$$h = (1-x)\ hf + x\ hg$$

$$v = (1-x)\ vf + x\ vg$$

$$s = (1-x)\ sf + x\ sg$$

where f and g indicate the property of the substance at saturated liquid and dry saturated vapour states respectively.

The heat supplied to the liquid for the complete phase change is called the specific enthalpy of vapourization.

$$hg - hf = hfg$$

STEAM TABLES

Water and steam properties play especially important roles in the steam power generation industry, where they are central not only to the design of boilers, turbines, and other equipment but also to the evaluation of equipment performance. For large-scale power generation, relatively small discrepancies in the calculation of key properties, such as the heat of vapourization, can translate into a difference of hundreds of thousands of dollars in the values calculated for the performance of boilers and turbines.

Such considerations make representations of the thermodynamic properties of water vital tools for process engineers. Characterizations of these properties are commonly known as "steam tables," though now they are more often in computerized form.

An International Formulating Committee was established in 1967, and the 1967 Formulation for Industrial Use, known as IFC-67, was produced. This internationally accepted formulation, best known to U.S. engineers from the ASME Steam Tables, has formed the basis for calculations in the power industry for the past 30 years.

While IFC-67 has been a success, it left room for improvement in accuracy, speed, and internal consistency. In the early 1990s, IAPWS collected information on the industrial use of IFC-67 and identified shortcomings. Simultaneously, IAPWS produced a new Formulation for General and Scientific Use, which was adopted in 1995 (called here IAPWS-95). This formulation provides a state-of-the-art, accurate representation of the thermodynamic properties of water and steam that is recommended for research use and general use outside the power industry where computational speed is not an issue. It was decided that a new industrial formulation should agree with IAPWS-95, within certain tolerances. These tolerances provide a balance among available physical accuracy, the limits of precision required in industrial use, and the strong industrial requirement for calculational speed. In 1997, the IAPWS Industrial Formulation 1997 (IAPWS-IF97, or just IF97) was formally adopted for international use.

RELATIONSHIP BETWEEN PRESSURE AND TEMPERATURE FOR SATURATED VAPOUR

The average energy of the particles in a liquid is governed by the temperature. The higher the temperature, the higher the average energy. But within that average, some particles have energies higher than the average, and others have energies lower than the average. Some of the more energetic particles on the surface of the liquid can be moving fast enough to escape from the attractive forces holding the liquid together. They evaporate.

The diagram shows a small region of a liquid near its surface.

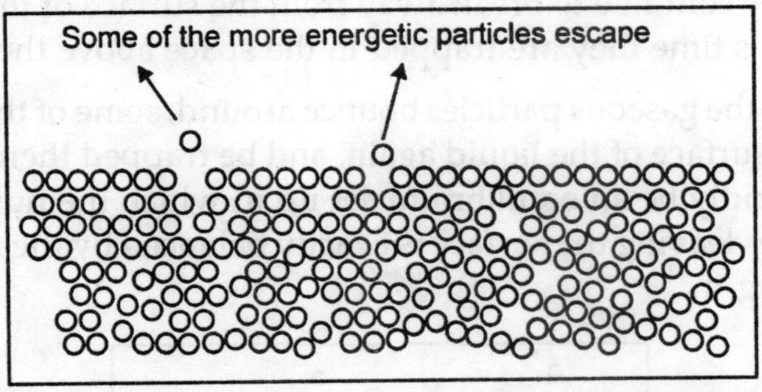

Notice that evaporation only takes place on the surface of the liquid. That's quite different from boiling which happens when there is enough energy to disrupt the attractive forces throughout the liquid. That's why, if you look at boiling water, you see bubbles of gas being formed all the way through the liquid. If you look at water which is just evaporating in the sun, you don't see any bubbles. Water molecules are simply breaking away from the surface layer.

Eventually, the water will all evaporate in this way. The energy which is lost as the particles evaporate is replaced from the surroundings. As the molecules in the water jostle with each other, new molecules will gain enough energy to escape from the surface.

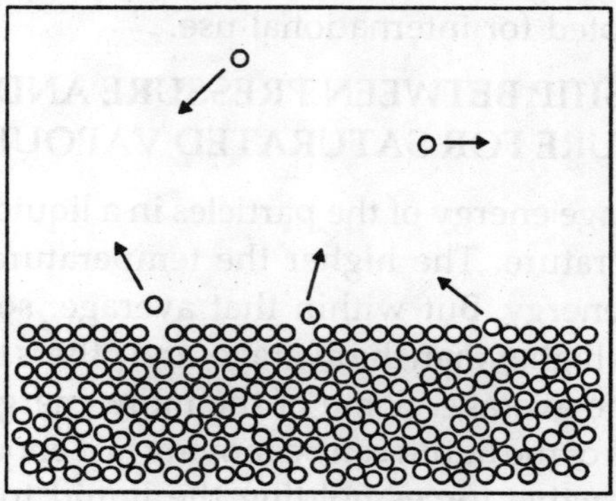

Now imagine what happens if the liquid is in a closed container. Common sense tells you that water in a sealed bottle doesn't seem to evaporate - or at least, it doesn't disappear over time. But there is constant evaporation from the surface. Particles continue to break away from the surface of the liquid - but this time they are trapped in the space above the liquid.

As the gaseous particles bounce around, some of them will hit the surface of the liquid again, and be trapped there. There will rapidly be an equilibrium set up in which the number of particles leaving the surface is exactly balanced by the number rejoining it.

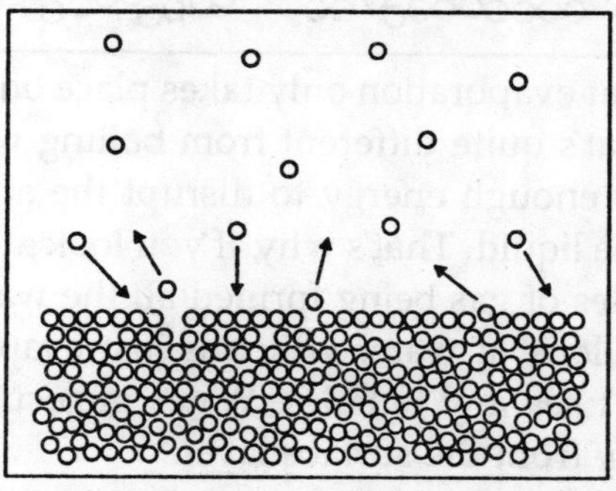

In this equilibrium, there will be a fixed number of the gaseous particles in the space above the liquid. When these particles hit the walls of the container, they exert a pressure. This pressure is called the saturated vapour pressure (also known as saturation vapour pressure) of the liquid.

Calculation of the Saturated Vapour Pressure

It isn't difficult to show the existence of this saturated vapour pressure (and to measure it) using a simple piece of apparatus.

If you have a mercury barometer tube in a trough of mercury, at 1 atmosphere pressure the column will be 760 mm tall. 1 atmosphere is sometimes quoted as 760 mmHg ("millimetres of mercury").

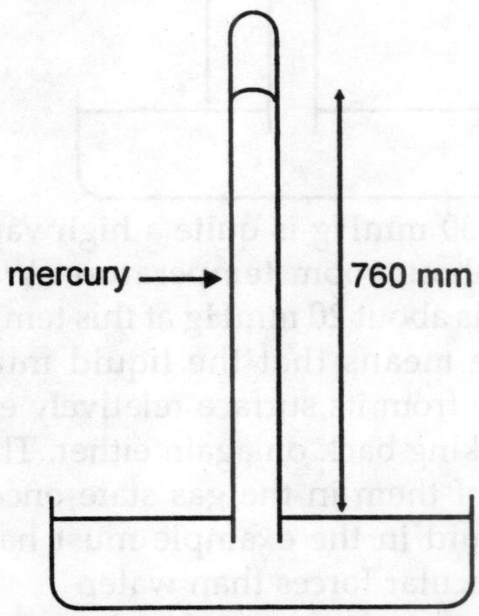

If you squirt a few drops of liquid into the tube, it will rise to form a thin layer floating on top of the mercury. Some of the liquid will evaporate and you will get the equilibrium we've just been talking about - provided there is still some liquid on top of the mercury. It is only an equilibrium if both liquid and vapour are present.

The saturated vapour pressure of the liquid will force the mercury level down a bit. You can measure the drop - and this gives a value for the saturated vapour pressure of the liquid at this temperature. In this case, the mercury has been forced down by a distance of 760 - 630 mm. The saturated vapour pressure of this liquid at the temperature of the experiment is 130 mmHg. You could convert this into proper SI units (pascals) if you wanted to. 760 mmHg is equivalent to 101325 Pa.

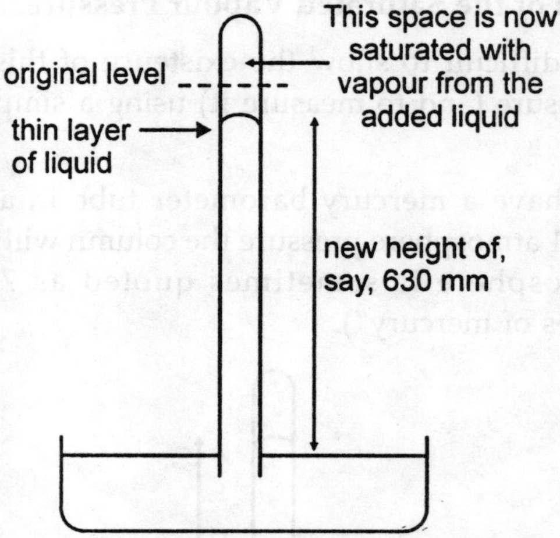

A value of 130 mmHg is quite a high vapour pressure if we are talking about room temperature. Water's saturated vapour pressure is about 20 mmHg at this temperature. A high vapour pressure means that the liquid must be volatile - molecules escape from its surface relatively easily, and aren't very good at sticking back on again either. That will result in larger numbers of them in the gas state once equilibrium is reached. The liquid in the example must have significantly weaker intermolecular forces than water.

The variation of Saturated Vapour Pressure with Temperature

The Effect of Temperature on the Equilibrium between Liquid and Vapour

If you increase the temperature, you are increasing the average energy of the particles present. That means that more of them are likely to have enough energy to escape from the surface of the liquid. That will tend to increase the saturated vapour pressure. Or you can look at it in terms of Le Chatelier's Principle - which works just as well in this kind of physical situation as it does in the more familiar chemical examples.

When the space above the liquid is saturated with vapour particles, you have this equilibrium occurring on the surface of the liquid:

Liquid ⇌ vapour ΔH is + ve

The forward change (liquid to vapour) is endothermic. It needs heat to convert the liquid into the vapour.

According to Le Chatelier, increasing the temperature of a system in a dynamic equilibrium favours the endothermic change. That means that increasing the temperature increases the amount of vapour present, and so increases the saturated vapour pressure.

The effect of temperature on the saturated vapour pressure of water

The graph shows how the saturated vapour pressure (svp) of water varies from 0°C to 100 °C. The pressure scale (the vertical one) is measured in kilopascals (kPa). 1 atmosphere pressure is 101.325 kPa.

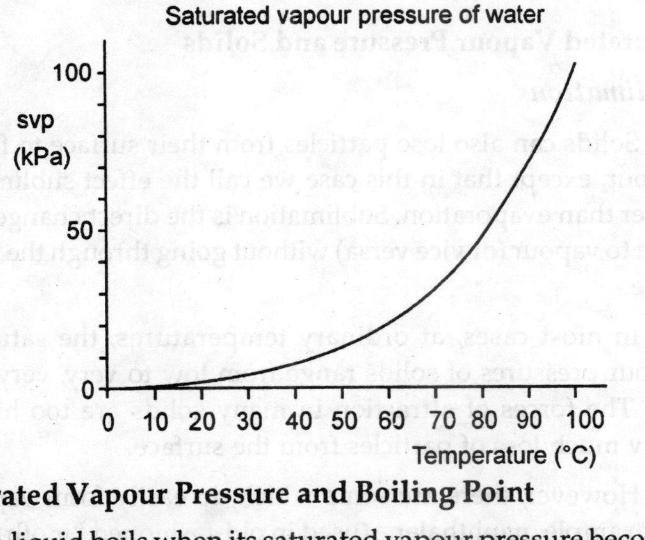

Saturated Vapour Pressure and Boiling Point

A liquid boils when its saturated vapour pressure becomes equal to the external pressure on the liquid. When that happens, it enables bubbles of vapour to form throughout the liquid - those are the bubbles you see when a liquid boils.

If the external pressure is higher than the saturated vapour pressure, these bubbles are prevented from forming, and you just get evaporation at the surface of the liquid.

If the liquid is in an open container and exposed to normal atmospheric pressure, the liquid boils when its saturated vapour pressure becomes equal to 1 atmosphere (or 101325 Pa or 101.325 kPa or 760 mmHg). This happens with water when the temperature reaches 100°C.

But at different pressures, water will boil at different temperatures. For example, at the top of Mount Everest the pressure is so low that water will boil at about 70°C. Depressions from the Atlantic can easily lower the atmospheric pressure in the UK enough so that water will boil at 99°C - even lower with very deep depressions.

Whenever we just talk about "the boiling point" of a liquid, we always assume that it is being measured at exactly 1 atmosphere pressure. In practice, of course, that is rarely exactly true.

Saturated Vapour Pressure and Solids

Sublimation

Solids can also lose particles from their surface to form a vapour, except that in this case we call the effect sublimation rather than evaporation. Sublimation is the direct change from solid to vapour (or vice versa) without going through the liquid stage.

In most cases, at ordinary temperatures, the saturated vapour pressures of solids range from low to very, very, very low. The forces of attraction in many solids are too high to allow much loss of particles from the surface.

However, there are some which do easily form vapours. For example, naphthalene (used in old-fashioned "moth balls" to deter clothes moths) has quite a strong smell. Molecules must be breaking away from the surface as a vapour, because otherwise you wouldn't be able to smell it.

Another fairly common example is solid carbon dioxide - "dry ice". This never forms a liquid at atmospheric pressure and always converts directly from solid to vapour. That's why it is known as dry ice.

PHASE DIAGRAMS OF PURE SUBSTANCES

The Elementary Phase Diagram

At its simplest, a phase can be just another term for solid, liquid or gas. If you have some ice floating in water, you have a solid phase present and a liquid phase. If there is air above the mixture, then that is another phase. But the term can be used more generally than this. For example, oil floating on water also consists of two phases - in this case, two liquid phases. If the oil and water are contained in a bucket, then the solid bucket is yet another phase. In fact, there might be more than one solid phase if the handle is attached separately to the bucket rather than moulded as a part of the bucket.

You can recognise the presence of the different phases because there is an obvious boundary between them - a boundary between the solid ice and the liquid water, for example, or the boundary between the two liquids.

Phase Diagrams

A phase diagram lets you work out exactly what phases are present at any given temperature and pressure. In the cases we'll be looking at on this page, the phases will simply be the solid, liquid or vapour (gas) states of a pure substance.

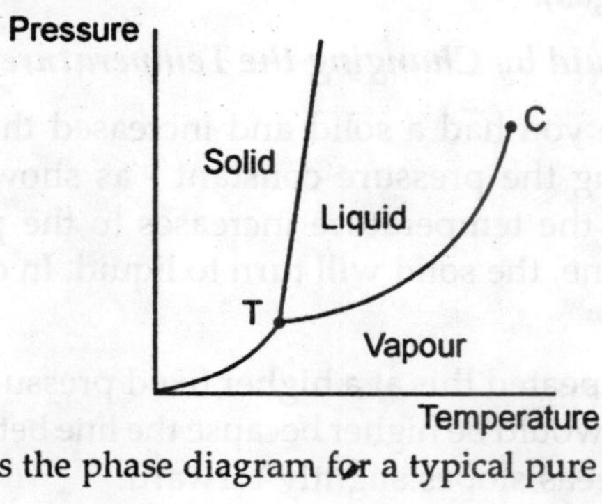

Fig. This is the phase diagram for a typical pure substance.

These diagrams (including this one) are nearly always drawn highly distorted in order to see what is going on more easily. There are usually two major distortions. We'll discuss these when they become relevant. If you look at the diagram, you will see that there are three lines, three areas marked "solid", "liquid" and "vapour", and two special points marked "C" and "T".

The Three Areas

Suppose you have a pure substance at three different sets of conditions of temperature and pressure corresponding to 1, 2 and 3 in the next diagram.

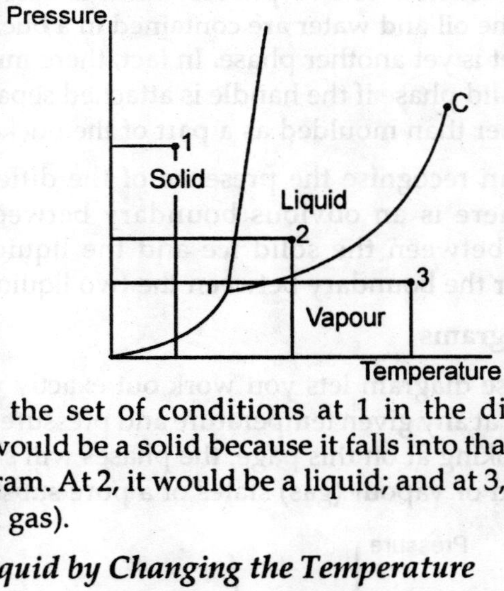

Under the set of conditions at 1 in the diagram, the substance would be a solid because it falls into that area of the phase diagram. At 2, it would be a liquid; and at 3, it would be a vapour (a gas).

Solid to Liquid by Changing the Temperature

Suppose you had a solid and increased the temperature while keeping the pressure constant - as shown in the next diagram. As the temperature increases to the point where it crosses the line, the solid will turn to liquid. In other words, it melts.

If you repeated this at a higher fixed pressure, the melting temperature would be higher because the line between the solid and liquid areas slopes slightly forward.

So what actually is this line separating the solid and liquid areas of the diagram? It simply shows the effect of pressure on melting point. Anywhere on this line, there is an equilibrium between solid and liquid.

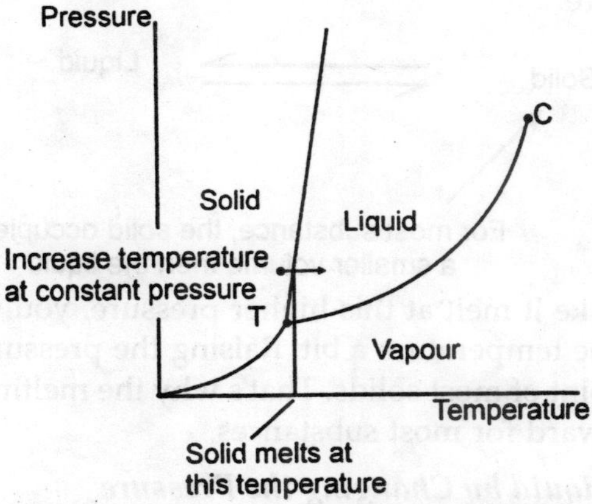

You can apply Le Chatelier's Principle to this equilibrium just as if it was a chemical equilibrium. If you increase the pressure, the equilibrium will move in such a way as to counter the change you have just made.

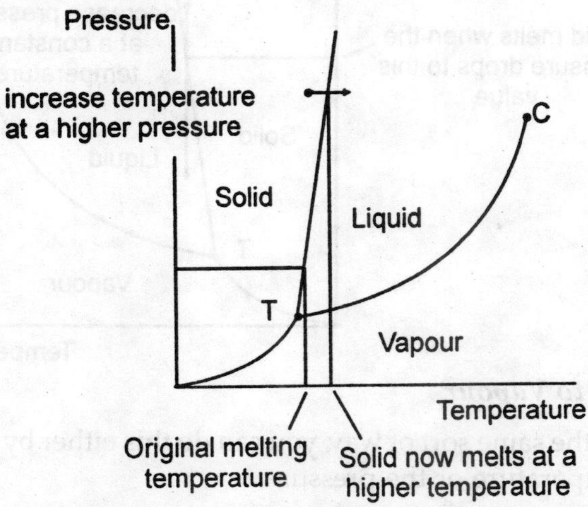

If it converted from liquid to solid, the pressure would tend to decrease again because the solid takes up slightly less space for most substances. That means that increasing the pressure on the equilibrium mixture of solid and liquid at its original melting point will convert the mixture back into the solid again. In other words, it will no longer melt at this temperature.

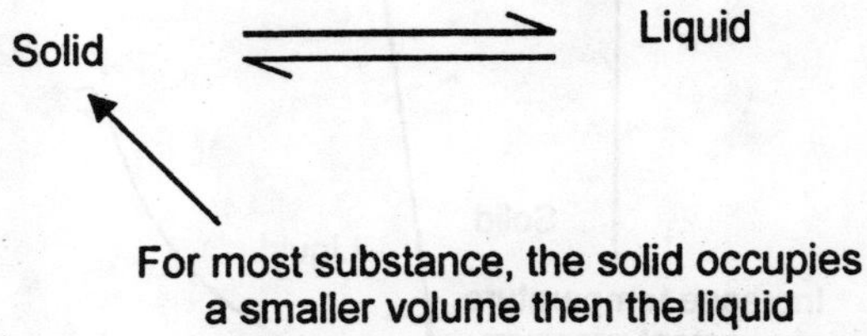

To make it melt at this higher pressure, you will have to increase the temperature a bit. Raising the pressure raises the melting point of most solids. That's why the melting point line slopes forward for most substances.

Solid to Liquid by Changing the Pressure

You can also see what happens if you decrease the pressure on a solid at constant temperature.

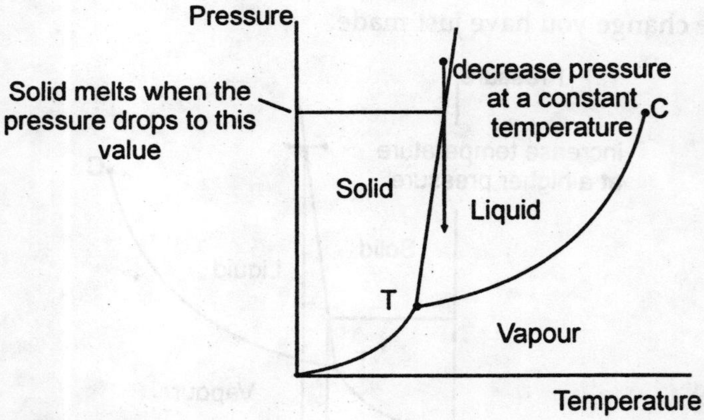

Liquid to Vapour

In the same sort of way, you can do this either by changing the temperature or the pressure.

The liquid will change to a vapour - it boils - when it crosses the boundary line between the two areas. If it is temperature that you are varying, you can easily read off the boiling temperature from the phase diagram. In the diagram above, it is the temperature where the red arrow crosses the boundary line.

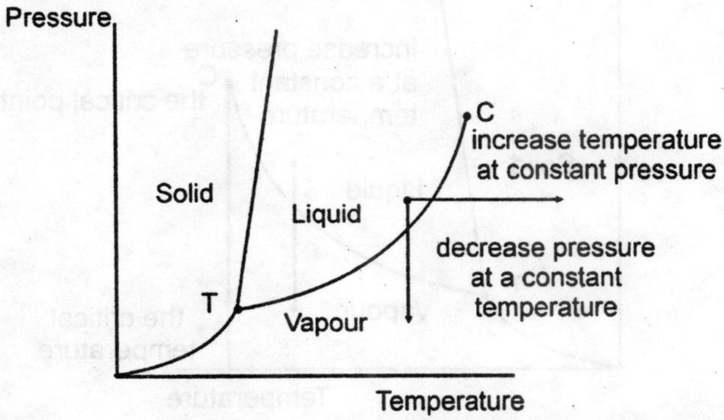

So, again, what is the significance of this line separating the two areas? Anywhere along this line, there will be an equilibrium between the liquid and the vapour. The line is most easily seen as the effect of pressure on the boiling point of the liquid. As the pressure increases, so the boiling point increases.

The Critical Point

You will have noticed that this liquid-vapour equilibrium curve has a top limit that we have labelled as **C** in the phase diagram.

This is known as the critical point. The temperature and pressure corresponding to this are known as the critical temperature and critical pressure.

If you increase the pressure on a gas (vapour) at a temperature lower than the critical temperature, you will eventually cross the liquid-vapour equilibrium line and the vapour will condense to give a liquid.

This works fine as long as the gas is below the critical temperature. What, though, if your temperature was above the critical temperature? There wouldn't be any line to cross.

That is because, above the critical temperature, it is impossible to condense a gas into a liquid just by increasing the pressure. All you get is a highly compressed gas. The particles have too much energy for the intermolecular attractions to hold them together as a liquid.

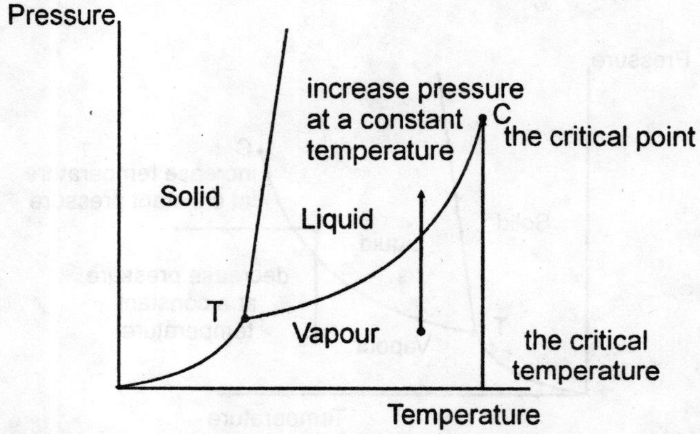

The critical temperature obviously varies from substance to substance and depends on the strength of the attractions between the particles. The stronger the intermolecular attractions, the higher the critical temperature.

Moving from Solid to Vapour

There's one more line to look at on the phase diagram. This is the line in the bottom left-hand corner between the solid and vapour areas.

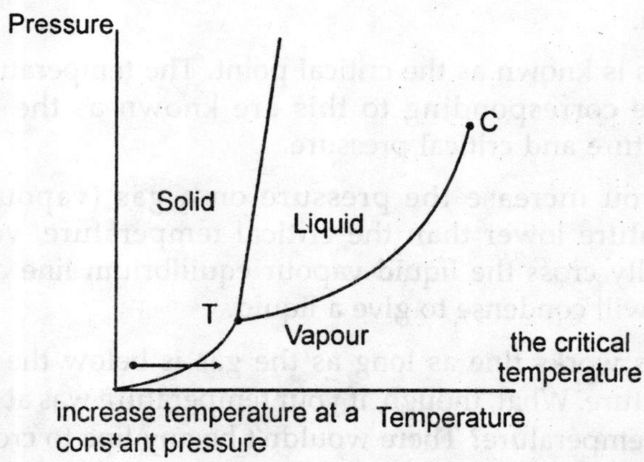

That line represents solid-vapour equilibrium. If the conditions of temperature and pressure fell exactly on that line, there would be solid and vapour in equilibrium with each other - the solid would be subliming. (Sublimation is the change directly from solid to vapour or vice versa without going through the liquid phase.)

We can cross that line by either increasing the temperature of the solid, or decreasing the pressure. The diagram shows the effect of increasing the temperature of a solid at a (probably very low) constant pressure. The pressure obviously has to be low enough that a liquid can't form - in other words, it has to happen below the point labelled as T.

You could read the sublimation temperature off the diagram. It will be the temperature at which the line is crossed.

The Triple Point

Point T on the diagram is called the triple point. If you think about the three lines which meet at that point, they represent conditions of:

- Solid-liquid equilibrium
- Liquid-vapour equilibrium
- Solid-vapour equilibrium

Where all three lines meet, you must have a unique combination of temperature and pressure where all three phases are in equilibrium together. That's why it is called a triple point. If you controlled the conditions of temperature and pressure in order to land on this point, you would see an equilibrium which involved the solid melting and subliming, and the liquid in contact with it boiling to produce a vapour - and all the reverse changes happening as well. If you held the temperature and pressure at those values, and kept the system closed so that nothing escaped, that's how it would stay. A strange set of affairs!

Normal Melting and Boiling Points

The normal melting and boiling points are those when the pressure is 1 atmosphere. These can be found from the phase diagram by drawing a line across at 1 atmosphere pressure.

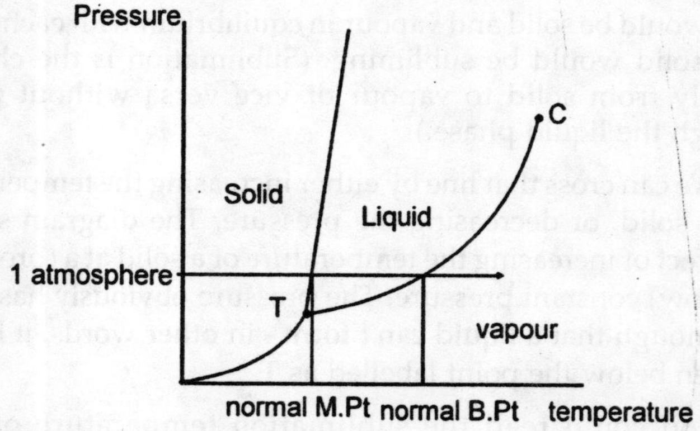

The Phase Diagram for Water

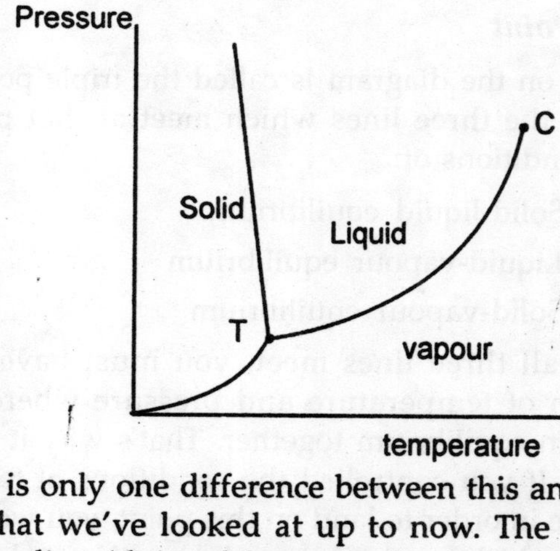

There is only one difference between this and the phase diagram that we've looked at up to now. The solid-liquid equilibrium line (the melting point line) slopes backwards rather than forwards.

In the case of water, the melting point gets lower at higher pressures.

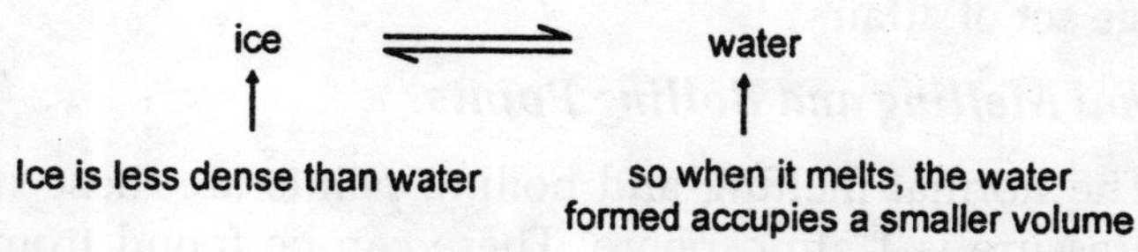

If you have this equilibrium and increase the pressure on it, according to Le Chatelier's Principle the equilibrium will move to reduce the pressure again. That means that it will move to the side with the smaller volume. Liquid water is produced.

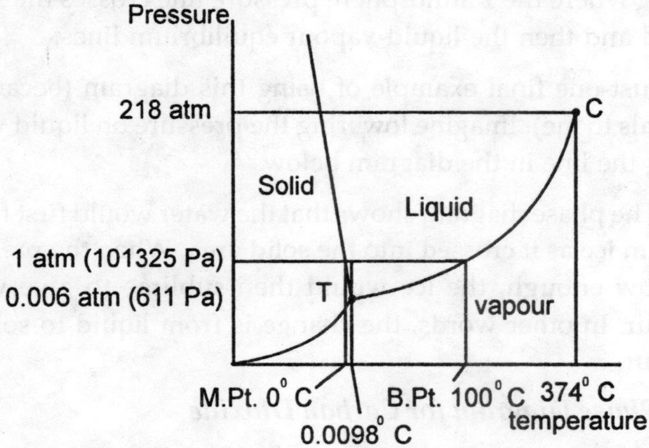

To make the liquid water freeze again at this higher pressure, you will have to reduce the temperature. Higher pressures mean lower melting (freezing) points. Now lets put some numbers on the diagram to show the exact positions of the critical point and triple point for water.

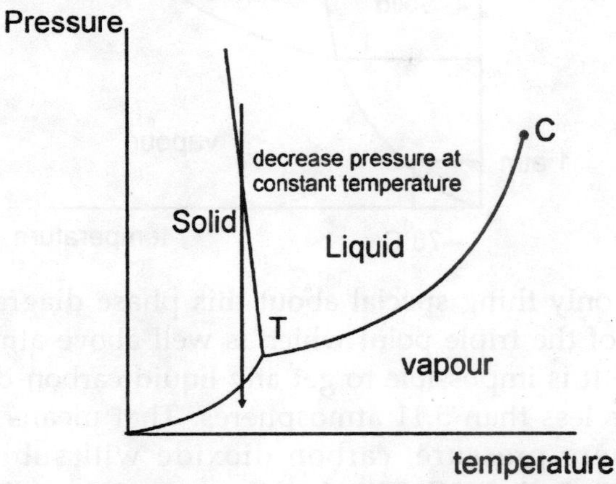

Notice that the triple point for water occurs at a very low pressure. Notice also that the critical temperature is 374°C. It would be impossible to convert water from a gas to a liquid by compressing it above this temperature.

The normal melting and boiling points of water are found in exactly the same way as we have already discussed - by seeing where the 1 atmosphere pressure line crosses the solid-liquid and then the liquid-vapour equilibrium lines.

Just one final example of using this diagram (because it appeals to me). Imagine lowering the pressure on liquid water along the line in the diagram below.

The phase diagram shows that the water would first freeze to form ice as it crossed into the solid area. When the pressure fell low enough, the ice would then sublime to give water vapour. In other words, the change is from liquid to solid to vapour.

The Phase Diagram for Carbon Dioxide

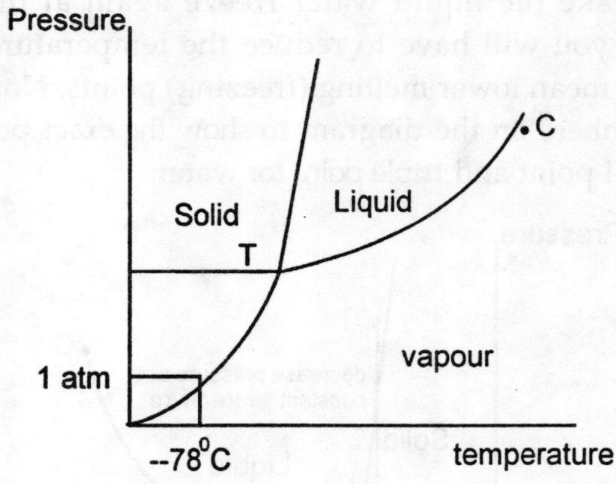

The only thing special about this phase diagram is the position of the triple point which is well above atmospheric pressure. It is impossible to get any liquid carbon dioxide at pressures less than 5.11 atmospheres. That means that at 1 atmosphere pressure, carbon dioxide will sublime at a temperature of -78°C. This is the reason that solid carbon dioxide is often known as "dry ice". You can't get liquid carbon dioxide under normal conditions - only the solid or the vapour.

Chapter 8

Ideal Gases

CRITICAL TEMPERATURE

The critical temperature of a liquid is that temperature beyond which the liquid cannot exist, no matter how much pressure is placed on it. The pressure that is needed to cause the gas to condense at the critical temperature is the critical pressure.

On a phase diagram, this point is noted by the ending of the liquid/vapour equilibrium line.

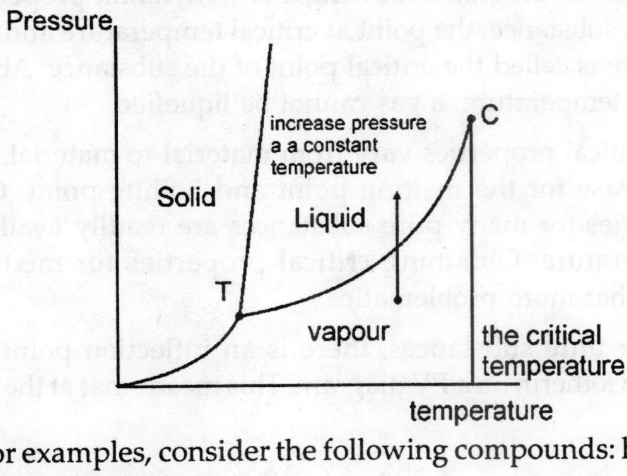

For examples, consider the following compounds: helium, propane and water.

Helium has a critical temperature of –268°C. Above this temperature, liquid helium cannot exist, no matter how much pressure is placed on it.

Propane, however, has a critical temperature of 97°C. This means it can exist as a liquid or as a gas below 97°C, depending on the external pressure. At room pressure, propane is a gas, but it can be easily liquified by placing it in a container and pressurizing it. This is how propane is frequently sold: in 5–25 pound tanks of pressurized, liquid propane at room temperature

Water has a critical temperature of 374°C. However, at room temperature and pressure, water is a liquid, unlike propane. If we reduce the pressure exerted on a sample of liquid water, we can get it to boil at room temperature

The critical temperature, T_c, of a material is the temperature above which distinct liquid and gas phases do not exist. As the critical temperature is approached, the properties of the gas and liquid phases become the same. Above the critical temperature, there is only one phase: that of supercritical fluid. The critical pressure is the vapour pressure at the critical temperature. The critical molar volume is the volume of one mole of material at the critical temperature and pressure. On diagrams showing thermodynamic properties for a given substance, the point at critical temperature and critical pressure is called the critical point of the substance. Above its critical temperature, a gas cannot be liquefied.

Critical properties vary from material to material, just as is the case for the melting point and boiling point. Critical properties for many pure substances are readily available in the literature. Obtaining critical properties for mixtures is somewhat more problematic.

For pure substances, there is an inflection point in the critical isotherm on a PV diagram. This means that at the critical point:

$$\left(\frac{\partial P}{\partial V}\right)_T = \left(\frac{\partial^2 P}{\partial V^2}\right)_T = 0$$

This relation can be used to evaluate two parameters for an equation of state in terms of the critical properties.

Sometimes a set of reduced properties are defined in terms of the critical properties, ie.:

$$T_r = T / T_c$$

$$P_r = P / P_c$$

$$V_r = V / V_c$$

The principle of corresponding states indicates that substances at equal reduced pressures and temperatures have equal reduced volumes. This relationship is approximately true for many substances, but becomes increasingly inaccurate for large values of P_r.

Two immiscible liquids, such as oil and water, will also have a critical temperature and pressure at which the two phases will become consolute.

Critical temperature of some substances

Element	**Critical Temperature (°C)**
Ammonia	132
Argon	–122.29
Arsenic	1400
Bromine	315
Carbon dioxide (CO_2)	31.2
Carbon monoxide (CO)	–141
Chlorine	143.8
Ethanol (C_2H_6O)	216
Fluorine	–128.85
Helium	–267.96
Hydrogen	–240.17
Iodine	546
Krypton	–63.7
Mercury	1477
Methane (CH_4)	–82
Neon	–228.75

Nitrogen	–146.9
Oxygen	–118.57
Phosphorus	721
Propane (C_3H_8)	97
Radon	105
Selenium	1493
Sulphur	1041
Sulphur dioxide (SO_2)	157
Water (H_2O)	374
Xenon	16.58

Critical Point

In thermodynamics, a critical point, also called a critical state, specifies the conditions (temperature, pressure) at which the liquid state of the matter ceases to exist. As a liquid is heated, its density decreases while the pressure and density of the vapour being formed increases. The liquid and vapour densities become closer and closer to each other until the critical temperature is reached where the two densities are equal and the liquid-gas line or phase boundary disappears. More generally, the critical point is the point of termination of a phase equilibrium curve, which separates two distinct phases. At this point, the phases are no longer distinguishable.

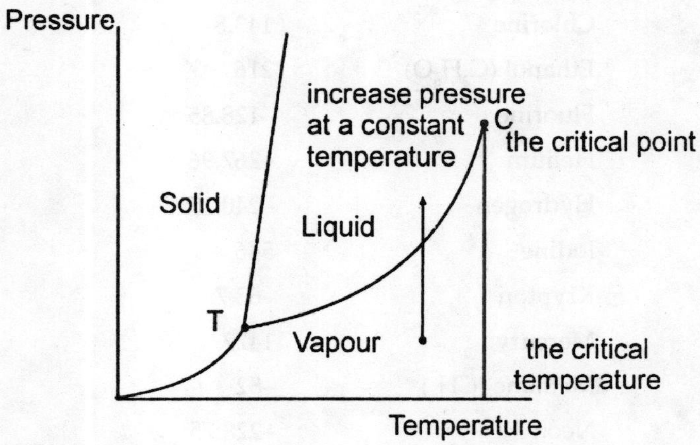

The critical point in a phase diagram is at the high-temperature extreme of the liquid-gas phase boundary.

In the phase diagram shown, the phase boundary between liquid and gas does not continue indefinitely. Instead, it terminates at a point on the phase diagram called the critical point. This reflects the fact that, at extremely high temperatures and pressures, the liquid and gaseous phases become indistinguishable. In water, the critical point occurs at around 647 K (374°C or 705°F) and 22.064 MPa (3200 PSIA or 218atm).

Critical variables are useful for rewriting a varied equation of state into one that applies to all materials. The effect is similar to a normalizing constant. According to renormalization group theory, the defining property of criticality is that the natural length scale characteristic of the structure of the physical system, the so-called correlation length becomes infinite. There are also lines in phase space along which this happens: these are critical lines.

In equilibrium systems the critical point is reached only by tuning a control parameter precisely. However, in some non-equilibrium systems the critical point is an attractor of the dynamics in a manner that is robust with respect to system parameters, a phenomenon referred to as self-organized criticality.

IDEAL GAS

A gas that, when kept at a constant temperature, would obey the gas laws exactly. No known gas is an ideal gas. An ideal gas or perfect gas is a gas consisting of identical particles of negligible volume, with no intermolecular forces. Additionally, the constituent atoms or molecules undergo perfectly elastic collisions with the walls of the container. The real gases that actually exist do not exhibit these exact properties, although the approximation is often good enough to treat real gases as ideal gases.

An ideal gas is one which behaves almost as a perfect gas, whose temperature is above the critical one and whose molecules have a simple monatomic structure.

What exactly is an ideal gas? An ideal gas is one that exactly conforms to the kinetic theory. The kinetic theory, as stated by Rudolf Clausius in 1857, has five key points. These are:

1. Gases are made of molecules in constant, random movement. Gases like Argon have 1-atom molecules.
2. The large portion of the volume of a gas is empty space. The volume of all gas molecules, in comparison, is negligible.
3. The molecules show no forces of attraction or repulsion.
4. No energy is lost in collision of molecules; the impacts are completely elastic.
5. The temperature of a gas is the average kinetic energy of all of the molecules.

The Kinetic Theory makes several assumptions about an ideal gas. These cause problems because real gases are not ideal. The main causes of error are related to pressure and temperature.

Pressure

At high pressures, the behaviour of real gases changes dramatically from that predicted by the Ideal Gas Law. Under 10 atmospheres of pressure or less, Ideal Gas Law predictions are very close to real amounts and do not generate serious error.

Temperature

When the temperature of a gas is close to its liquefaction point, the behaviour is very different from Ideal Gas Law predictions. With increasing temperatures, the Ideal Gas Law predictions become close to real values.

In order to overcome the errors in the Ideal Gas Law, Johannes van der Waals developed an equation to predict the behaviour of real gases. Johannes van der Waals' equation included corrections for the finite volume of the molecules of the gas and the attractive forces between the molecules. Two new constants, a and b, were added. The corrected equation is:

$$P = (nRT)/(V-nb) - (n^2a)/(V^2)$$

The correction nb subtracts the volume of the molecules. b is measured in liters/mole. The correction with a reflects the strength of attraction and is measured in liters^2-atmosperes per moles^2.

The equation is generally put in the form:

$$\left(P+\frac{n^2a}{V^2}\right)(V\text{-}nb) = n\,R\,T$$

Values of a and b are different for each gas. The values of a and b generally increase with increased mass of the molecule and complexity of the molecule.

BOYLE'S LAW

Boyle's Law states that under conditions of constant temperature and quantity, there is an inverse relationship between the volume and pressure for an ideal gas. Boyle's law is a basic law in chemistry describing the behaviour of an ideal gas under a constant temperature. An ideal gas is a perfectly pure gas undergoing perfect elastic collisions with its container. No such gas actually exists, but real gases behave closely enough to ideal gases that we can use theories applying to the latter to describe the former.

Boyle's law states that the volume of a gas is inversely proportional to the pressure applied to it. In other words, the pressure of a gas times its volume is a constant. It is not necessary to know the exact value of this constant to meaningfully understand the behaviour of gases. The law was discovered by Robert Boyle in 1662. The standard measurements for volume and pressure are cubic meters and pascals (or atmospheres).

One obvious example of Boyle's law in action is in a syringe. In a syringe, the volume of a fixed amount of gas is increased by drawing the handle back, thereby lessening the pressure. The blood in a vein has higher pressure than the gas in the syringe, so it flows into the syringe, equalizing the pressure differential.

Boyle's law is one of three gas laws which thoroughly describe the behaviour of gases under varying temperatures, pressures and volumes. The other two laws are Gay-Lussac's law and Graham's law.

Another way of describing Boyle's law is that when you push a gas, it tends to push back. Without the massive amount of gravity holding them together, the solar system's gas planets would rapidly diffuse in all directions, quickly depressurizing. There is a limit to the amount we can compress any given gas sample, because eventually the pressure becomes so great that it bursts out of any container we can create for it.

Boyle's Law states that the product of the volume and pressure of a fixed quantity of an ideal gas is constant, given constant temperature. Expressed mathematically, the formula for Boyle's law is:

$$PV = K$$

where:

- V is volume of the gas.
- P is the pressure of the gas.
- K is a constant.

The value of K is computed from measurements of volume and pressure for a fixed quantity of gas. After making a change to the system, typically by forcing a change in the volume of the vessel containing the fixed quantity of gas, the new volume and new pressure are measured. The result of computing the product of the measured new volume and the new pressure should be the original value of the constant K. Equation says that, after forcing the volume V of the fixed quantity of gas to increase, keeping the gas at the initially measured temperature, the pressure P must decrease proportionally. Conversely, reducing the volume of the gas increases the pressure.

Boyle's law is commonly used to predict the result of introducing a change, in volume and pressure only, to the initial state of a fixed quantity of gas. The "before" and "after" volumes and pressures of the fixed amount of gas, where the "before" and "after" temperatures are the same (heating or cooling will be required to meet this condition), are related by the equation:

$$P_1V_1 = P_2V_2$$

In practice, this equation is solved for one of the two "after" quantities to determine the effect that a change in the other "after" quantity will have. For example:

$$P_2 = P_1V_1/V_2$$

Boyle's law, Charles's Law, and Gay-Lussac's Law form the combined gas law. The three gas laws in combination with Avogadro's Law can be generalized by the ideal gas law.

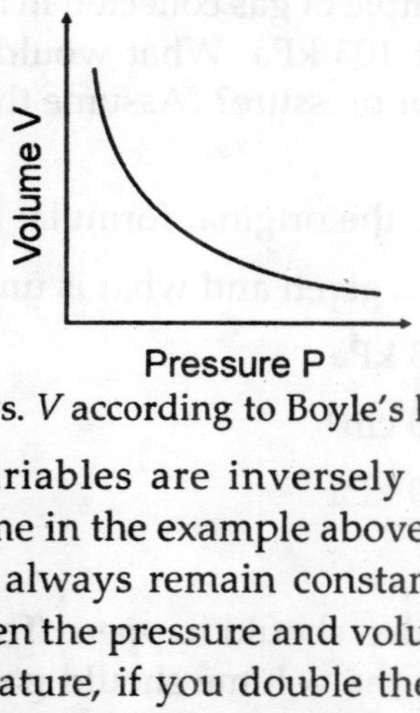

Fig. A graph of *P* vs. *V* according to Boyle's law is shown above

When two variables are inversely proportional, like pressure and volume in the example above, the product of the two variables will always remain constant. Because of the relationship between the pressure and volume of a gas sample at constant temperature, if you double the value of one, you divide the other by two. The chart below will demonstrate the inverse relationship between the volume and pressure of a gas. Imagine a gas sample trapped in a cylinder which allows you to adjust the pressure. Notice how the pressure changes cause the volume to change, while the product of the two variables will remain a constant (K).

Data for a Sample of Gas at Constant Temperature and Varying Pressure

Trial	Pressure	Volume	Formula	Calculation
1	100 kPa	40 cm^3	PV=K	100 kPa x 40 cm^3 = 4000 kPa x cm^3
2	50 kPa	80 cm^3	PV=K	50 kPa x 80 cm^3 = 4000 kPa x cm^3
3	200 kPa	20 cm^3	PV=K	200 kPa x 20 cm^3 = 4000 kPa x cm^3
4	400 kPa	10 cm^3	PV=K	400 kPa x 10 cm^3 = 4000 kPa x cm^3
5	25 kPa	160 cm^3	PV=K	25 kPa x 160 cm^3 = 4000 kPa x cm^3

Boyle's law is sometimes used to determine the volume that a gas would have at another pressure. If you were to collect a sample of gas under the atmospheric conditions in your lab on a given day, you might want to mathematically determine what the vclume of the gas would be under different conditions. The formula that can be used to calculate the affects of pressure changes on the volume of a gas at constant temperature is shown below:

$$P_1V_1 = P_2V_2$$

Example: A sample of gas collected in a 350 cm^3 container exerts a pressure of 103 kPa. What would be the volume of this gas at 150 kPa of pressure? (Assume that the temperature remains constant.)

Solution: Write the original formula: $P_1V_1 = P_2V_2$

Then list what is given and what is unknown.

$P_1 = 103$ kPa

$V_1 = 350$ cm^3

$P_2 = 150$ kPa

$V_2 = ?$

Next, predict what should happen. The pressure is going up by nearly 1/3, so the volume should go down by a bit less than 1/3.

Now, Adjust the original formula to isolate the unknown, solve and round to the correct number of significant digits.

a. $P_1V_1 = P_2V_2$

b. $P_1V_1 = P_2V_2$

c. $\frac{V_2 \cancel{P_2}}{\cancel{P_2}} = \frac{P_1V_1}{P_2}$

$P_1 = 103\ \text{kPa}$ $V_2 = \frac{103\ \text{kPa} \times 350\ \text{cm}^3}{150\ \text{kPa}}$

$V_1 = 350\ \text{cm}^3$

$P_2 = 150\ \text{kPa}$ $V_2 = 240.333333\ \text{cm}^3$

$V_2 = ?$ $V_2 = 240\ \text{cm}^3$

Of course, the original formula can be solved for a different unknown. For example, you can determine what the pressure would have to be in order to end up with a certain volume.

Example: A sample of neon has a volume of 239 cm³ at 2.00 atm of pressure. What would the pressure have to be in order for the gas to have a volume of $5.00 \times 10^2\ \text{cm}^3$?

Solution: Write the original formula: $P_1V_1 = P_2V_2$

Then list what is given and what is unknown.

$P_1 = 2.00\ \text{atm}$

$V_1 = 239\ \text{cm}^3$

$P_2 = ?$

$V_2 = 5.00 \times 10^2\ \text{cm}^3$

Next, predict what should happen. You want the volume to more than double, so the pressure would have to be less than half the original.

Now, Adjust the original formula to isolate the unknown, solve and round to the correct number of significant digits.

a. $P_1V_1 = P_2V_2$

b. $\frac{P_1V_1}{V_2} = \frac{P_2\ V_2}{V_2}$

c. $P_2 = \dfrac{P_1V_1}{V_2}$

$P_1 = 2.00$ atm	$P_2 = \dfrac{2.00 \text{ atm} \times 239 \text{ cm}^3}{5.00 \times 10^2 \text{ cm}^3}$
$V_1 = 239$ cm^3	
$P_2 = ?$	$P_2 = 0.956$ atm
$V_2 = 5.00 \times 10^2$ cm^3	$P_2 = 0.956$ atm

Solve

- If some neon gas at 121 kPa were allowed to expand from 3.7 dm^3 to 6.0 dm^3 without changing the temperature, what pressure would the neon gas exert under these new conditions?
- A quantity of gas under a pressure of 1.78 atm has a volume of 550 cm^3. The pressure is increased to 2.50 atm, while the pressure remains constant. What is the new volume?
- Under a pressure of 172 kPa, a gas has a volume of 564 cm^3. The pressure is decreased, without changing the temperature, until the volume of the gas is equal to 8.00×10^2 cm^3. What is the new pressure?
- A sample of gas under a pressure of 622 kPa has a volume of 233 cm^3. The pressure is increased to 988 kPa, while the temperature remains constant. What volume will the gas occupy at this pressure?
- The pressure exerted by a gas is 2.0 atm while it has a volume of 350 cm^3. What would be the volume of this sample of gas at standard atmospheric pressure?
- A sample of CO_2 has a volume of 2.7 dm^3 at 78.5 kPa. At what pressure would this sample of gas have a volume of 4.0 dm^3?

CHARLES'S LAW

Jacques Charles investigated the relationship between the Volume of a gas and how it changes with temperature. He noted that the volume of a gas increased with the temperature. Charles's Law states that the volume of a given amount of dry ideal gas is directly proportional to the Kelvin Temperature provided the amount of gas and the pressure remain fixed. When we plot the Volume of a gas against the Kelvin temperature it forms a straight line. The mathematical statement is that the V / T = a constant. For two sets of conditions the following is a math statement of Charles's Law:

$$V_1 / T_1 = V_2 / T_2$$

An example of Charles's Law would be what happens when a hot air balloon has air heated. The air expands and fills the balloon. Of course, other physical principles cause the balloon to rise against the gravitational force. As the air inside the balloon expands the balloon gets bigger and displaces more air. The displaced air produces a buoyant force that counters the gravitational force and causes the balloon to rise.

This law gives the relationship between volume and temperature if pressure and amount are held constant.

If the volume of a container is increased, the temperature increases.

If the volume of a container is decreased, the temperature decreases.

Suppose the temperature is increased. This means gas molecules will move faster and they will impact the container walls more often. This means the gas pressure inside the container will increase (but only for an instant. The greater pressure on the inside of the container walls will push them outward, thus increasing the volume. When this happens, the gas molecules will now have farther to go, thereby lowering the number of impacts and dropping the pressure back to its constant value.

It is important to note that this momentary increase in pressure lasts for only a very, very small fraction of a second. You would need a very fast, accurate pressure sensing device to measure this momentary change. Consider another case. Suppose the volume is suddenly increased. This will reduce the pressure, since molecules now have farther to go to impact the walls. However, this is not allowed by the law; the pressure must remain constant. Therefore, the temperature must go up, in order to get the molecules to the walls faster, thereby overcoming the longer distance and keeping the pressure constant.

Charles' Law is a direct mathematical relationship. This means there are two connected values and when one goes up, the other also increases.

The mathematical form of Charles' Law is:

$$V/T = K$$

This means that the volume-temperature fraction will always be the same value if the pressure and amount remain constant. Let V_1 and T_1 be a volume-temperature pair of data at the start of an experiment. If the volume is changed to a new value called V_2, then the temperature must change to T_2.

The new volume-temperature data pair will preserve the value of k. The ChemTeam does not care what the actual value of k is, only that two different volume-temperature data pairs equal the same value and that value is called k.

So we know

$$V_1/T_1 = K$$

$$V_2/T_2 = K$$

we can conclude that

$$V1/T1 = V2/T2.$$

$$\frac{V_1}{T_1} = \frac{V_2}{T_2} \qquad V_1T_2 = V_2T_1$$

Notice that the right-hand equation results from cross-multiplying the first one.

Note: Every temperature used in a calculation must be in Kelvin, not degrees Celsius.

Don't use Celsius in a numerical calculation. Use Kelvin every time.

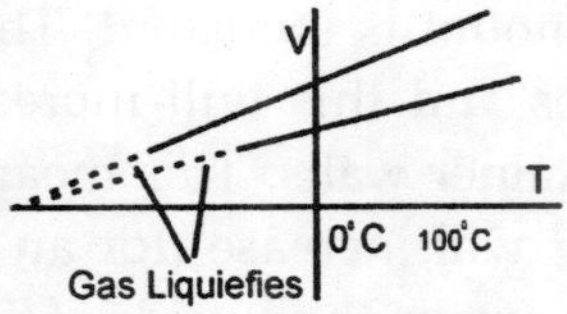

Fig. A graph of *V* vs. *T* according to Charles law is shown above

Example: A gas is collected and found to fill 2.85 L at 25.0°C. What will be its volume at standard temperature?

Solution: Convert 25.0°C to Kelvin and you get 298 K. Standard temperature is 273 K. We use equation like this:

$$\frac{2.85\text{ L}}{298\text{ K}} = \frac{x}{273\text{ K}}$$

Example: 4.40 L of a gas is collected at 50.0°C. What will be its volume upon cooling to 25.0°C?

Solution: Convert 50.0°C to 323 K and 25.0°C to 298 K. Then plug into the equation and solve for x, like this:

$$\frac{4.40\text{ L}}{323\text{ K}} = \frac{x}{298\text{ K}}$$

Example: 5.00 L of a gas is collected at 100 K and then allowed to expand to 20.0 L. What must the new temperature be in order to maintain the same pressure (as required by Charles' Law)?

Solution:

$$\frac{5.00\text{ L}}{100\text{ K}} = \frac{20.0\text{ L}}{x}$$

AVOGADRO'S LAW

Discovered by Amedo Avogadro, this law gives the relationship between volume and amount when pressure and temperature are held constant. Remember amount is measured in moles.

If the amount of gas in a container is increased, the volume increases. If the amount of gas in a container is decreased, the volume decreases. Let us discuss why?

Suppose the amount is increased. This means there are more gas molecules and this will increase the number of impacts on the container walls. This means the gas pressure inside the container will increase (for an instant), becoming greater than the pressure on the outside of the walls. This causes the walls to move outward. Since there is more wall space the impacts will lessen and the pressure will return to its original value.

The mathematical form of Avogadro's Law is:

$$V/n = K$$

This means that the volume-amount fraction will always be the same value if the pressure and temperature remain constant.

Let V_1 and n_1 be a volume-amount pair of data at the start of an experiment. If the amount is changed to a new value called n_2, then the volume will change to V_2.

We know

$$V_1/n_1 = K$$

$$V_2/n_2 = K$$

We can conclude

$$V_1/n_1 = V_2/n_2.$$

$$\boxed{\frac{V_1}{n_1} = \frac{V_2}{n_2} \qquad V_1 n_2 = V_2 n_1}$$

Example: 5.00 L of a gas is known to contain 0.965 mol. If the amount of gas is increased to 1.80 mol, what new volume will result (at an unchanged temperature and pressure)?

Solution:

Using $V_1 n_2 = V_2 n_1$

$$(5.00 \text{ L})\ (1.80 \text{ mol}) = (x)\ (0.965 \text{ mol})$$

IDEAL GAS EQUATION

The three historically important gas laws derived relationships between two physical properties of a gas, while keeping other properties constant:

Boyle's laq: $V \propto \frac{1}{P}$ (constant n, T)

Charles's law: $V \propto T$ (constant n, P)

Avogadro's law: $V \propto n$ (constant P, T)

These different relationships can be combined into a single relationship to make a more general gas law:

$$V \propto \frac{nT}{P}$$

If the proportionality constant is called "R", then we have:

$$V = R\left(\frac{nT}{P}\right)$$

Rearranging to a more familiar form:

$$PV = nRT$$

This Equation is Known as the Ideal-gas Equation

- An "ideal gas" is one whose physical behaviour is accurately described by the ideal-gas equation
- The constant R is called the gas constant
  - The value and units of R depend on the units used in determining P, V, n and T
  - Temperature, T, must always be expressed on an absolute-temperature scale (K)
  - The quantity of gas, n, is normally expressed in *moles*
  - The units chosen for pressure and volume are typically *atmospheres* (*atm*) and *liters* (*l*), however, other units may be chosen
  - *PV* can have the units of energy:

The Gas Constant *R*

Repeated experiments show that at standard temperature (273 K) and pressure (1 atm or 101325 N/m^2), one mole (n = 1) of gas occupies 22.4 L volume. Using this experimental value, you can evaluate the gas constant ***R***,

$$R = \frac{PV}{nT} = \frac{1\,\text{atm } 22.4\,\text{L}}{1\,\text{mol } 273\,\text{K}} = 0.08205\,\text{L atm/(mol.K)}$$

When SI units are desirable, P = 101325 N/m^2 (Pa for pascal) instead of 1 atm. The volume is 0.0224 m^3. The numberical value and units for R is

$$R = \frac{101325\ \text{N/m}^2\ 0.0224\ \text{m}^3}{1\ \text{mol}\ \ 273\ \text{K}}$$

$$= 8.314\ \text{J/(mol.K)}$$

Note that 1 L atm = 0.001 m^3 × 101325 N / m^2 = 101.325 J (or N m). Since energy can be expressed in many units, other numerical values and units for R are frequently in use.

The gas constant R is such a universal constant for all gases that its values are usually listed in the "Physical Constants" of textbooks and handbooks. Although we try to use SI units all the time, the use of atm for pressure is still common.

Thus, we often use R = 8.314 J / (mol·K) or 8.3145 J / mol·K.

Gas constant can be expressed in the following values and units

Values for the gas constant *R*

Units	*Value*
L atm/mol K	0.08206
cal/mol K	1.987
J/mol K	8.314
m^3 Pa/mol K	8.314
L torr/mol K	62.36

Example: If we had 1.0 mol of gas at 1.0 atm of pressure at 0°C (273.15 K), what would be the volume?

Solution:

PV = nRT

V = nRT/P

V = (1.0 mol)(0.0821 L atm/mol K)(273 K)/(1.0 atm)

V = 22.41 L

Applications of the Ideal Gas Law

The ideal gas law has four parameters and a constant R.

$$\mathbf{P\,V = n\,R\,T}$$

and it can be rearranged to give an expression for each of P, V, n or T. For example,

$$P = n\,R\,T\,/\,V, \quad \text{(Boyles law)}$$

$$P = (n\,R\,/\,V)\,T \quad \text{(Charles law)}$$

These equations are Boyles law and Charles law respectively. Similar expressions can be derived for V, n and T in terms of other variables. Thus, there are many applications. However, you must make sure that you use the proper numerical value for the gas constant R according to the units you have for the parameters.

Furthermore, n / V is number of moles per unit volume, and this quantity has the same units as the concentration (C). Thus, the concentration is a function of pressure and temperature,

$$C = P\,/\,R\,T$$

At 1.0 atm pressure and room temperature of 298 K, the concentration of an ideal gas is 0.041 mol/L.

The Avogadro's law can be further applied to correlate gas density d (weight per unit volume or n M / V) and molecular mass M of a gas. The following equation is easily derived from the ideal gas law:

$$PM = \frac{nM}{V}RT$$

Thus, we have

$$P M = d R T / M$$

$$d = n M / V$$

$$d = P M / R T$$

$$M = d R T / P$$

Example: An air sample containing only nitrogen and oxygen gases has a density of 1.3393 g / L at STP. Find the weight and mole percentages of nitrogen and oxygen in the sample.

Solution: From the density d, we can evaluate an average molecular weight (also called molar mass).

$P M = d R T$

$M = 22.4 \times d$

$= 22.4$ L/mol $\times$ 1.3393 g/L

$= 30.0$ g / mol

Assume that we have 1.0 mol of gas, and x mol of which is nitrogen, then (1 – x) is the amount of oxygen. The average molar mass is the mole weighted average, and thus,

$28.0\,x + 32.0\,(1 - x) = 30.0$

$-4\,x = -2$

$x = 0.50$ mol of N_2, and $1.0 - 0.50 = 0.50$ mol O_2

Now, to find the weight percentage, find the amounts of nitrogen and oxygen in 1.0 mol (30.0 g) of the mixture.

Mass of 0.5 mol nitrogen = $0.5 \times 28.0 = 14.0$ g

Mass of 0.5 mol oxygen = $0.5 \times 32.0 = 16.0$ g

Percentage of nitrogen = $100 \times 14.0 / 30.0 = 46.7$ per cent

Percentage of oxygen = $100 \times 16.0 / 30.0 = 100 - 46.7 = 53.3$ per cent

We can find the density of pure nitrogen and oxygen first and evaluate the fraction from the density.

d of N_2 = 28.0 / 22.4 = 1.2500 g/L

d of O_2 = 32.0 / 22.4 = 1.4286 g/L

1.2500 x + 1.4286 (1 – x) = 1.3393

Solving for x gives

x = 0.50 (same result as above)

Example: What is the density of acetone, C_3H_6O, vapour at 1.0 atm and 400 K?

Solution: The molar mass of acetone = 3×12.0 + 6 ×1.0 + 16.0 = 58.0. Thus,

d = P M / R T

= (1.0 × 58.0 atm g/mol) / (0.08205 L atm / (mol K) × 400 K)

= 1.767 g / L

NON-IDEAL GASES

An ideal gas is a theoretical idea – a gas in which there are no attractive forces between the molecules, and in which the molecules take up no space. Both of these assumptions are incorrect.

- If there were no attractive forces between molecules, no substances would ever condense from the gas state to become liquids and solids. In fact, every substance *does* condense when it is cooled and compressed enough. Therefore there *must* be attractive forces between molecules.
- If molecules took up no space, then we could compress substances until they had no volume at all. In fact, solids and liquids are almost incompressible, since their molecules are basically touching each other. Since solids and liquids take up space, the molecules of which they are made *must* take up space also.

However:

- When the pressure is fairly low (around atmospheric pressure or less) the molecules are not very close to each other. The intermolecular attractions and molecular size are not very important under these conditions.
- When the temperature is fairly high (perhaps 100ºC or more above the substance's boiling boint) the attractive forces are not very important compared to the rapid motion of the molecules

Under these conditions, a gas will behave nearly ideally. This is true for things we readily identify as gases at room temperature and pressure conditions, like hydrogen, nitrogen and oxygen. These real gases are said to behave ideally – that is, they obey the ideal gas law.

Most of us are familiar with a disposable butane lighter. Under a moderate amount of pressure, the butane is in the liquid state at room temperature. We know that the pressure can't be too high, because the lighter is made of fairly thin plastic that would not stand a high pressure. When the pressure is released, the butane liquid vapourizes, and becomes a gas, since the boiling point of iso-butane is –11.7 ºC, approximately 35 degrees below room temperature. These characteristics:

- Low pressure required to condense the gas
- Boiling point close to room temperature suggest that iso-butane probably deviates a lot from ideal gas behaviour.

Chapter 9

Thermodynamic Processes

A thermodynamic process can be defined as "an operation during which the properties of state, pressure, volume and temperature may change, with energy transfer in the form of work and/or heat flow taking place"

Whenever one or more of the properties of a system change, a change in the state of the system occurs. The path of the succession of states through which the system passes is called the thermodynamic process. One example of a thermodynamic process is increasing the temperature of a fluid while maintaining a constant pressure. Another example is increasing the pressure of a confined gas while maintaining a constant temperature.

SECOND LAW OF THERMODYNAMICS

One of the earliest statements of the Second Law of Thermodynamics was made by R. Clausius in 1850. He stated

> "It is impossible to construct a device that operates in a cycle and produces no effect other than the removal of heat from a body at one temperature and the absorption of an equal quantity of heat by a body at a higher temperature".

The second law states that there exists a useful state variable called entropy S. The change in entropy delta S is equal to the heat transfer delta Q divided by the temperature T.

$$\text{delta } S = \text{delta } Q / T$$

For a given physical process, the combined entropy of the system and the environment remains a constant if the process can be reversed. If we denote the initial and final states of the system by "i" and "f":

Sf = Si (reversible process)

An example of a reversible process is ideally forcing a flow through a constricted pipe. Ideal means no boundary layer losses. As the flow moves through the constriction, the pressure, temperature and velocity change, but these variables return to their original values downstream of the constriction. The state of the gas returns to its original conditions and the change of entropy of the system is zero. Engineers call such a process an isentropic process. Isentropic means constant entropy.

The second law states that if the physical process is irreversible, the combined entropy of the system and the environment must increase. The final entropy must be greater than the initial entropy for an irreversible process:

Sf > Si (irreversible process)

In a general sense, the Second Law says that the differences between systems in contact with each other tend to even out. Pressure differences, density differences, and particularly temperature differences, all tend to equalize if given the opportunity. This means that an isolated system will eventually come to have a uniform temperature. A thermodynamic engine is an engine that provides useful work from the difference in temperature of two bodies. Since any thermodynamic engine requires such a temperature difference, it follows that no useful work can be derived from an isolated system in equilibrium, there must always be energy fed from the outside. The Second Law is often invoked as the reason why perpetual motion machines cannot exist.

The Second Law can be stated in various succinct ways, including:

- It is impossible to produce work in the surroundings using a cyclic process connected to a single heat reservoir (Kelvin, 1851).

- If thermodynamic work is to be done at a finite rate, free energy must be expended.

A mathematical statement of the Second Law is:

$$\frac{dS}{dt} \geq 0 \text{ on average}$$

where

S is the entropy and

t is time.

Note that this equation does not *need* to be true, but the Second Law asserts that it is *usually* true. Entropy can decrease, but this simply tends not to happen as often. A common misconception is that the Second Law means that entropy never decreases. In fact, the Second Law asserts only a statistical tendency, hence it is only *highly unlikely* that entropy will decrease in a closed system at any given instant.

Entropy

One consequence of the second law is the development of the physical property of matter termed entropy (S). Entropy was introduced to help explain the Second Law of Thermodynamics. The change in this property is used to determine the direction in which a given process will proceed. Entropy can also be explained as a measure of the unavailability of heat to perform work in a cycle. This relates to the second law since the second law predicts that not all heat provided to a cycle can be transformed into an equal amount of work, some heat rejection must take place. The change in entropy is defined as the ratio of heat transferred during a reversible process to the absolute temperature of the system. In the early 1850s, thermodynamicist Rudolf Clausius began to put the thermodynamic concepts of energy dissipation on a mathematical footing. Clausius defined the change in entropy dS of a thermodynamic system, during a reversible process, as

$$DS = \delta Q/T_{abs} \text{ (For a reversible process)}$$

where

S = The change in entropy of a system during some process (Btu/°R)

δ Q = The amount of heat added to the system during the process (Btu)

D T_{abs} = The absolute temperature at which the heat was transferred (°R)

This expression comes from the Clausius theorem. This definition makes sense when absolute temperature has been defined. Note that the small amount δ*Q* of energy transferred by heating is denoted by δ*Q* rather than *dQ*, because *Q* is not a state function while the entropy is.

Clausius gave the quantity *S* the name "entropy", from the Greek word *τροπή*, "transformation". Since this definition involves only differences in entropy, the entropy itself is only defined up to an arbitrary additive constant.

When a process is irreversible, the above definition must be replaced by the statement that the entropy change is equal to the amount of energy required to return the system to its original state by a reversible transformation at a constant temperature, divided by that temperature.

Clausius' identification of *S* as a significant quantity was motivated by the study of reversible and irreversible thermodynamic transformations. A thermodynamic transformation is a change in a system's thermodynamic properties, such as temperature and volume. A transformation is reversible if it is quasistatic which means that it is infinitesimally close to thermodynamic equilibrium at all times. Otherwise, the transformation is irreversible. To illustrate this, consider a gas enclosed in a piston chamber, whose volume may be changed by moving the piston. If we move the piston slowly enough, the density of the gas is always homogeneous, so the transformation is reversible. If we move the piston quickly, pressure waves are created, so the gas is not in equilibrium, and the transformation is irreversible.

A heat engine is a thermodynamic system that can undergo a sequence of transformations which ultimately return it to its original state. Such a sequence is called a cyclic process, or simply a *cycle*. During some transformations, the engine may exchange energy with the environment. The net result of a cycle is

1. Mechanical work done by the system (which can be positive or negative, the latter meaning that work is done *on* the engine), and
2. Heat energy transferred from one part of the environment to another. By the conservation of energy, the net energy lost by the environment is equal to the work done by the engine.

If every transformation in the cycle is reversible, the cycle is reversible, and it can be run in reverse, so that the energy transfers occur in the opposite direction and the amount of work done switches sign.

Temperature

In thermodynamics, absolute temperature is *defined* in the following way. Suppose we have two heat reservoirs, which are systems sufficiently large that their temperatures do not change when energy flows into or out of them. A reversible cycle exchanges heat with the two heat reservoirs:

$$\frac{Q_1}{T_1} = \frac{Q_2}{T_2}$$

where

Q_1 is the amount of heat that the cycle absorbs from the first reservoir,

T_1 is the temperature of the first reservoir and

Q_2 is the amount of heat delivered to the second,

T_2 is the temperature of the second reservoir.

Introduce an additional heat reservoir at an arbitrary temperature T_0, as well as N cycles with the following property: the j-th such cycle operates between the T_0 reservoir and the T_j reservoir, transferring energy δQ_j to the latter. From the above definition of temperature, the energy extracted from the T_0 reservoir by the j-th cycle is

$$\delta Q_{0,j} = T_0 \frac{\delta Q_j}{T_j}$$

Now consider one cycle of the heat engine, accompanied by one cycle of each of the smaller cycles. At the end of this process, each of the N reservoirs have zero net energy loss (since the energy extracted by the engine is replaced by the smaller cycles), and the heat engine has done an amount of work equal to the energy extracted from the T_0 reservoir,

$$W = \sum_{j=1}^{N} \delta Q_{0,j} = T_0 \sum_{j=1}^{N} \frac{\delta Qj}{T_j}$$

If this quantity is positive, this process would be a perpetual motion machine of the second kind, which is impossible. Thus,

$$\sum_{i=1}^{N} \frac{\delta Q_i}{T_i} \leq 0$$

Now repeat the above argument for the reverse cycle. The result is

$$\sum_{i=1}^{N} \frac{\delta Q_i}{T_i} = 0 \quad \text{(reversible cycles)}$$

Now consider a reversible cycle in which the engine exchanges heats δQ_1, δQ_2,..., δQ_N with a sequence of N heat reservoirs with temperatures T_1,..., T_N. We can show that:

$$\sum_{i=1}^{N} \frac{\delta Q_i}{T_i} = 0$$

where

δQ_i is the amount of heat exchanged *from* the engine *to* the reservoir,

T_i is the temperature of each heat reservoir, and

i is an integer representing each different heat reservoir.

Since the cycle is reversible, the engine is always infinitesimally close to equilibrium, so its temperature is equal to any reservoir with which it is contact. In the limiting case of a reversible cycle consisting of a *continuous* sequence of transformations,

$$\oint \frac{\delta Q}{T} = 0 \text{ (reversible cycles)}$$

where the integral is taken over the entire cycle, and T is the temperature of the system at each point in the cycle. This is a particular case of the Clausius theorem.

State Function

We can now deduce an important fact about the entropy change during *any* thermodynamic transformation, not just a cycle. First, consider a reversible transformation that brings a system from an equilibrium state A to another equilibrium state B. If we follow this with *any* reversible transformation which returns that system to state A, our above result says that the net entropy change is zero. This implies that the entropy change in the first transformation depends *only on the initial and final states*.

This allows us to define the entropy of any *equilibrium* state of a system. Choose a reference state R and call its entropy S_R. The entropy of any equilibrium state X is

$$S_X = S_R + \int_R^X \frac{\delta Q}{T}$$

Since the integral is independent of the particular transformation taken, this equation is well-defined.

Entropy Changes in Irreversible Processes

We now consider irreversible Processes. It can be shown that the entropy change during any transformation between two *equilibrium* states is

$$\Delta S \geq \int \frac{\delta Q}{T}$$

where the equality holds if the transformation is reversible.

Notice that if $\delta Q = 0$, then ΔS e ≥ 0. This is the Second Law of Thermodynamics, which we have discussed earlier.

Suppose a system is thermally and mechanically isolated from the environment. For example, consider an insulating rigid box divided by a movable partition into two volumes, each filled with gas. If the pressure of one gas is higher, it will expand by moving the partition, thus performing work on the other gas. Also, if the gases are at different temperatures, heat can flow from one gas to the other provided the partition is an imperfect insulator. Our above result indicates that the entropy of the system *as a whole* will increase during these process (it could in principle remain constant, but this is unlikely.) Typically, there exists a maximum amount of entropy the system may possess under the circumstances. This entropy corresponds to a state of *stable equilibrium*, since a transformation to any other equilibrium state would cause the entropy to decrease, which is forbidden. Once the system reaches this maximum-entropy state, no part of the system can perform work on any other part. It is in this sense that entropy is a measure of the energy in a system that "cannot be used to do work".

Effect of Temperature, Volume, and Concentration on the Entropy

As discussed previously, the entropy of a substance always increases with temperature. This is due mainly to the increased number of ways that the thermal energy can be distributed amongst the allowed energy levels as the latter become accessible. The rate of increase dS/dT is just the ratio of the heat capacity to the temperature C/T. When integrated over a range of temperatures, this yields (for an ideal gas confined to a fixed volume)

$$\Delta S = C \ln\left(\frac{T_2}{T_1}\right)$$

In general, a larger volume also leads to increased entropy. For an ideal gas that expands at a constant temperature (meaning that it absorbs heat from the surroundings to compensate for the work it does during the expansion), the increase in entropy is given by

$$\Delta S = R \ln\left(\frac{V_2}{V_1}\right)$$

Because the pressure of a gas is inversely proportional to its volume, we can easily alter the above relation to express the entropy change associated with a change in the pressure of a perfect gas:

$$\Delta S = R \ln\left(\frac{P_2}{P_1}\right)$$

The pressure of a gas is directly proportional to its concentration in moles per liter, so we can re-cast this equation in terms of concentrations:

$$\Delta S = R \ln\left(\frac{C_2}{C_1}\right)$$

Although these equations strictly apply only to perfect gases and cannot be used at all for liquids and solids, it turns out that in a dilute solution, the solute can often be treated as a gas dispersed in the volume of the solution, so the last equation can actually give a fairly accurate value for the entropy of dilution of a solution.

ENTROPY OF MIXING (IDEAL GASES)

Visualize that we have a container divided into two compartments. In one compartment we have n_1 moles of an ideal gas, gas 1, at pressure, p and temperature, T. In the other compartment we have n_2 moles of another ideal gas, gas 2, at the same p and T.

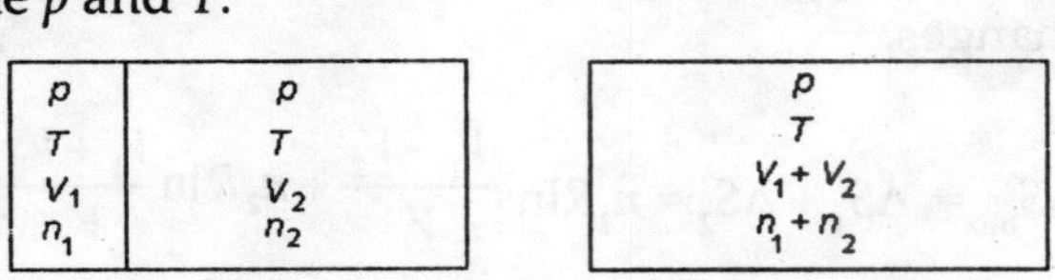

If we remove the partition the gases will begin to diffuse into each other and the system will eventually reach the state where both gases are uniformly distributed throughout the container. This is clearly an irreversibly process so that we would expect that the entropy would increase.

To calculate the entropy change we must find a reversible path to carry out the process, even if the path is fictitious. Imagine that we can devise a process that will expand one gas reversibly and isothermally, but leave the other gas undisturbed. We know how to calculate the change in entropy for the reversible isothermal expansion of an ideal gas.

Recall that $dU = 0$ for the isothermal expansion of an ideal gas. Then,

$$dU = TdS - pdV = 0$$

So,

$$TdS = pdV$$

$$dS = \frac{P}{T} dV$$

$$= \frac{nR}{V} dV,$$

for an ideal gas. So, for gas number 1 in our fictitious isothermal expansion we have,

$$\Delta S_1 = n_1 R \ln \frac{V_1 + V_2}{V_1},$$

and for gas number 2,

$$\Delta S_2 = n_2 R \ln \frac{V_1 + V_2}{V_2},$$

The total entropy change is the sum of these two individual entropy changes,

$$\Delta S_{mix} = \Delta S_1 + \Delta S_2 = n_1 R \ln \frac{V_1 - V_2}{V_1} + n_2 R \ln \frac{V_1 + V_2}{V_2}$$

Equation 4 could be used for calculations, but it is not in the form that we will see in other contexts. To obtain the usual form factor the R out of Equation 4 and invert the argument of the logarithms,

$$\Delta S_{\text{mix}} = -R\left(n_1 \ln \frac{V_1}{V_1 + V_2} + n_2 \ln \frac{V_2}{V_1 + V_2}\right)$$

(ince we are doing a calculation for ideal gases, notice that the argument of the first logarithm in Equation 5 can be written,

$$\frac{V_1}{V_1 + V_2} = \frac{\frac{n_1 RT}{p}}{\frac{n_1 RT}{p} + \frac{n_2 RT}{p}} = \frac{n_1}{n_1 + n_2} = X_1$$

where X_1 is the mole fraction of component 1. There is an equivalent expression for the argument of the second logarithm. The entropy of mixing becomes,

$$\Delta S_{\text{mix}} = -R(n_1 \ln X_1 + n_2 \ln X_2).$$

Equation 7 is also suitable for calculations, but it is not yet in the standard form. To obtain the standard form write the total number of moles $n_1 + n_2$ as n and multiply and divide equation by n. The result is

$$\Delta S_{\text{mix}} = -nR(X_1 \ln X_1 + X_2 \ln X_2).$$

Equation 8 is the same form that we will find when we derive an expression for the entropy of mixing ideal solutions and it is the same form that Shannon found for the "entropy of a message" in his famous series of papers on information theory.

If the two gases were not at the same initial pressure you would have to introduce some extra steps. Expand or compress one of the gases to bring it to the pressure of the other gas, mix the gases, and then compress or expand the mixture to bring it to the correct final volume and pressure.

If the two gases are not at the same temperature and pressure it is more complicated. You must find the final temperature using heat balance, reversibly cool and heat the two gases respectively to the same temperature, expand or contract on of the gases, mix them, and then expand or contract the mixture to the appropriate volume.

Equation 8 can easily be extended to more than two gases,

$$\Delta S_{mix} = -nR(X_1 \ln X_1 + X_2 \ln X_2 + X_3 \ln X_3 + ...)$$

Example - The molar entropy of dry air

The composition of dry air is approximately 78per cent N_2, 21per cent O_2, and 1per cent Ar by volume (which is the same as mole percent). What is the molar entropy of mixing of air?

$$\Delta S_{mix,air} = 1 \text{ mol} \times 8.314 \frac{\text{J}}{\text{K mol}} (0.78 \ln 0.78 + 0.21 \ln 0.21 + 0.01 \ln 0.01)$$

$$= 4.7 \frac{\text{J}}{\text{K}}.$$

What Does Entropy Measure

Entropy measures disorder.

If we look at the processes we have seen which have positive entropy changes we can see that in each case an increase in entropy is associated with an increase in disorder.

An isothermal expansion gives the molecules more room to move around in, the molecules are less localized.

Increasing the temperature increases the average speeds of the molecules. The molecules are said to be more disordered in "velocity space" (or momentum space).

Mixing gases (or liquids) intersperses the molecules among each other increasing the disorder.

Phase changes, such as going from a solid to a liquid or a gas, or from a liquid to a gas, increase the entropy because gases are more disordered than solids or liquids and liquids are more disordered than solids.

For example, the entropy of fusion of 1.00 mol of ice at 273.15 (heat of fusion is 6.008 kJ/mol) is

$$\Delta S_{\text{fus}} = \frac{6008\,\text{J}}{273.15\,\text{K}} = 22.0\frac{\text{J}}{\text{K}}.$$

Vapourization of liquids has a large positive entropy of vapourization because gases are greatly disordered compared to liquids. A typical value is obtained from the vapourization of benzene at its boiling point. The heat of vapourization is 30.8 kJ at the boiling point, 353.1K. The entropy of vapourization is

$$\Delta S_{\text{vap}} = \frac{30800\,\text{J}}{353.1\,\text{K}} = 87.2\frac{\text{J}}{\text{K}}.$$

It is interesting that the entropy of vapourization of many substances at their boiling points is close to about 86 J/K. (Water and helium are exceptions.) This phenomenon is called "Trouton's rule." It is easily understood on the basis of entropy being a measure of disorder. The vapourization process essentially "creates" a mole of disordered molecules (the gas) from a mole of highly ordered molecules (the solid or liquid). The gases are all at one atmosphere pressure because we are at their normal boiling point.

Measurement of Entropy

In real experiments, it is quite difficult to measure the entropy of a system. The techniques for doing so are based on the thermodynamic definition of the entropy, and require extremely careful calorimetry.

For simplicity, we will examine a mechanical system, whose thermodynamic state may be specified by its volume V and pressure P. In order to measure the entropy of a specific state, we must first measure the heat capacity at constant volume and at constant pressure (denoted C_V and C_P respectively), for a successive set of states intermediate between a reference state and the desired state. The heat capacities are related to the entropy S and the temperature T by

$$C_X = T\left(\frac{\partial S}{\partial T}\right)_X$$

where the X subscript refers to either constant volume or constant pressure. This may be integrated numerically to obtain a change in entropy:

$$\Delta S = \int \frac{C_X}{T} dT$$

We can thus obtain the entropy of any state (P,V) with respect to a reference state (P_0, V_0). The exact formula depends on our choice of intermediate states. For example, if the reference state has the same pressure as the final state,

$$S(P,V) = S(P,V_0) + \int_{T(P,V_0)}^{T(P,V)} \frac{C_P(P,V(T,P))}{T} dT$$

In addition, if the path between the reference and final states lies across any first order phase transition, the latent heat associated with the transition must be taken into account.

The entropy of the reference state must be determined independently. Ideally, one chooses a reference state at an extremely high temperature, at which the system exists as a gas. The entropy in such a state would be that of a classical ideal gas plus contributions from molecular rotations and vibrations, which may be determined spectroscopically. Choosing a *low* temperature reference state is sometimes problematic since the entropy at low temperatures may behave in unexpected ways. For instance, a calculation of the entropy of ice by the latter method, assuming no entropy at zero temperature, falls short of the value obtained with a high-temperature reference state by 3.41 J/(mol·K). This is due to the "zero-point" entropy of ice mentioned earlier.

Available Useful Work

An important and revealing idealised special case is to consider applying the Second Law to the scenario of an isolated system (called the total system or universe), made up of two parts: a sub-system of interest, and the sub-system's surroundings. These surroundings are imagined to be so large that they can be considered as an *unlimited* heat reservoir at temperature T_R and pressure P_R — so that no matter how much heat is transferred to (or from) the sub-system, the temperature of the surroundings will remain T_R; and no matter how much the volume of the sub-system expands (or contracts), the pressure of the surroundings will remain P_R.

Whatever changes dS and dS_R occur in the entropies of the sub-system and the surroundings individually, according to the Second Law the entropy S_{tot} of the isolated total system must increase:

$$dS_{tot} = dS + dS_R \geq 0$$

According to the First Law of Thermodynamics, the change dU in the internal energy of the sub-system is the sum of the heat δq added to the sub-system, *less* any work δw done *by* the sub-system, *plus* any net chemical energy entering the sub-system $\Sigma\mu_{iR}N_i$, so that:

$$dU = \delta q - \delta w + d\,(\Sigma\mu_{iR}N_i)$$

where μ_{iR} are the chemical potentials of chemical species in the external surroundings.

Now the heat leaving the reservoir and entering the sub-system is

$$\delta q = T_R(-dS_R) \leq T_R dS$$

where we have first used the definition of entropy in classical thermodynamics (alternatively, the definition of temperature in statistical thermodynamics); and then the Second Law inequality from above.

It therefore follows that any net work δw done by the sub-system must obey

$$\delta w \leq -dU + T_R dS + \Sigma\mu_{iR}dN_i$$

It is useful to separate the work done δw done by the subsystem into the *useful* work δw_u that can be done *by* the sub-system, over and beyond the work $P_R\,dV$ done merely by the sub-system expanding against the surrounding external pressure, giving the following relation for the useful work that can be done:

$$\delta w_u \leq -d(U - T_R S + P_R V - \Sigma\mu_{iR} N_i)$$

It is convenient to define the right-hand-side as the exact derivative of a thermodynamic potential, called the *availability* or exergy X of the subsystem,

$$X = U - T_R S + P_R V - \Sigma\mu_{iR} N_i$$

The Second Law therefore implies that for any process which can be considered as divided simply into a subsystem, and an unlimited temperature and pressure reservoir with which it is in contact,

$$dX + \delta w_u \leq 0$$

i.e. the change in the subsystem's exergy plus the useful work done *by* the subsystem (or, the change in the subsystem's exergy less any work, additional to that done by the pressure reservoir, done *on* the system) must be less than or equal to zero.

Various types of thermodynamic processes are:

Cyclic Process

When a system in a given initial state goes through a number of different changes in state (going through various processes) and finally returns to its initial values, the system has undergone a cyclic process or cycle. Therefore, at the conclusion of a cycle, all the properties have the same value they had at the beginning.

Steam (water) that circulates through a closed cooling loop undergoes a cycle.

Reversible Process

A reversible process for a system is defined as a process that, once having taken place, can be reversed, and in so doing leaves no change in either the system or surroundings. In other words the system and surroundings are returned to their original condition before the process took place. In reality, there are no truly reversible processes; however, for analysis purposes, one uses reversible to make the analysis simpler, and to determine maximum theoretical efficiencies. Therefore, the reversible process is an appropriate starting point on which to base engineering study and calculation. Although the reversible process can be approximated, it can never be matched by real processes. One way to make real processes approximate reversible process is to carry out the process in a series of small or infinitesimal steps. For example, heat transfer may be considered reversible if it occurs due to a small temperature difference between the system and its surroundings. For example, transferring heat across a temperature difference of 0.00001 °F "appears" to be more reversible than for transferring heat across a temperature difference of 100 °F.

Therefore, by cooling or heating the system in a number of infinitesamally small steps, we can approximate a reversible process. Although not practical for real processes, this method is beneficial for thermodynamic studies since the rate at which processes occur is not important.

Irreversible Process

An irreversible process is a process that cannot return both the system and the surroundings to their original conditions. That is, the system and the surroundings would not return to their original conditions if the process was reversed. For example, an automobile engine does not give back the fuel it took to drive up a hill as it coasts back down the hill.

There are many factors that make a process irreversible. Four of the most common causes of irreversibility are friction, unrestrained expansion of a fluid, heat transfer through a finite temperature difference, and mixing of two different substances. These factors are present in real, irreversible processes and prevent these processes from being reversible.

Adiabatic Process

An adiabatic process is one in which there is no heat transfer into or out of the system. The system can be considered to be perfectly insulated.

Example:

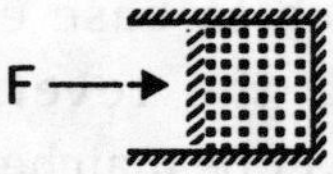

Q = 0

Compression of a Gas in an Insulated Cylinder.

In this case any change in the internal energy of the gas is due to work done on it or by it, $\Delta U = W$.

Normally if ΔU changes the temperature of a system will change. Any temperature rise or fall is due to the work done or by the gas alone and not due to heat flowing into or out of the system since $Q = 0$.

If a process is carried out fast enough the heat flow will be small and the process can be approximate as being adiabatic. This happen because heat flow is in general a slow process.

Observe that we did not say that Q is constant because it not a state variable. Q represent an energy transfer not the heat energy of the system.

In addition to the ideal gas law $PV = NkT$, the quantity PV^{γ} is constant for an ideal gas where $\gamma = c_P/c_V$, the ratio of molar specific heats.

For an ideal gas the work $W = (P_1V_1 - P_2V_2)/(\gamma - 1)$

Isoentropic Process

An isoentropic process is one in which the entropy of the fluid remains constant. This will be true if the process the system goes through is reversible and adiabatic. An isoentropic process can also be called a constant entropy process.

Example: A Heat Engine in which the working fluid undergoes an Adiabatic Reversible cyclic process.

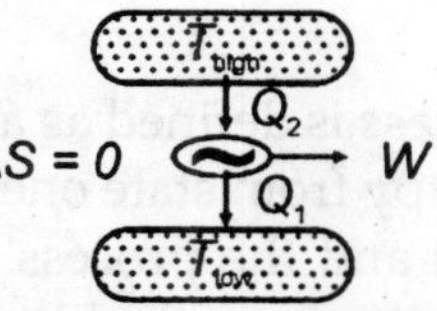

Any isoentropic process is also adiabatic since ΔU = ∨ dQ/T and Q = 0. However, not all adiabatic process are isoentropic. Adiabatic free expansion is not isoentropic.

For a heat engine to be reversible, not only must the change in entropy of the working fluid be zero but also the net change of the entropy of the environment (heat reservoirs) must also be zero.

Polytropic Process

When a gas undergoes a reversible process in which there is heat transfer, the process frequently takes place in such a manner that a plot of the Log P (pressure) vs. Log V (volume) is a straight line. Or stated in equation form PV^n = a constant. This type of process is called a polytropic process. An example of a polytropic process is the expansion of the combustion gasses in the cylinder of a water-cooled reciprocating engine.

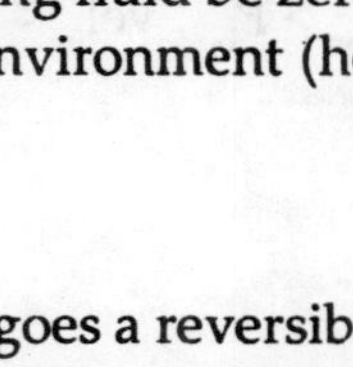

$\Delta(PV^n) = 0$

Here n can be any real number. All the process above can thought of as a special case of a polytropic process for an ideal gas.

n = 0 for Isobaric process since $PV^0 = P$ = constant.

n = 1 for Isothermal process since $PV^1 = PV = NKT$ = constant.

n = ∞ for Isovolumetric process

n = γ for Adiabatic process

Throttling Process

A throttling process is defined as a process in which there is no change in enthalpy from state one to state two, $h_1 = h_2$; no work is done, W = 0; and the process is adiabatic, Q = 0. To better understand the theory of the ideal throttling process let's compare what we can observe with the above theoretical assumptions. An example of a throttling process is an ideal gas flowing through a valve in midposition. From experience we can observe that:

$$P_{in} > P_{out}, \; vel_{in} < vel_{out}$$

where

P = pressure

vel = velocity

These observations confirm the theory that

$$h_{in} = h_{out}$$

Remember h = u + Pv (v = specific volume), so if pressure decreases then specific volume must increase if enthalpy is to remain constant (assuming u is constant). Because mass flow is constant, the change in specific volume is observed as an increase in gas velocity, and this is verified by our observations. The theory also states W = 0. Our observations again confirm this to be true as clearly no "work" has been done by the throttling process. Finally, the theory states that an ideal throttling process is adiabatic. This cannot clearly be proven by observation since a "real" throttling process is not ideal and will have some heat transfer.

Carnot's Principle

An irreversible heat engine operating between two heat reservoirs at constant temperatures cannot have an efficiency greater than that of a reversible heat engine operating between the same two temperatures.

Another form of the Second Law of Thermodynamics or Carnot's Principle is that: It is not possible to make a heat engine whose only effect is to absorb heat from a high-temperature region and turn all that heat into work. That is, it is not possible to design a heat engine that does not exhaust heat to the environment. Or, it is not possible to design a heat engine that has an efficiency of 1.00 or 100per cent.

Carnot Cycle

The Carnot cycle is the most efficient cycle possible. It consists of four basic reversible processes meaning that the cycle as a whole is also reversible. The four reversible processes are:

Adiabatic compression from T_2 to T_1.

Isothermal expansion (by placing the system in contact with a heat reservoir with temperature T_1).

Adiabatic expansion to $T_2 < T_1$.

Isothermal compression (by placing the system in contact with a heat reservoir with temperature T_2).

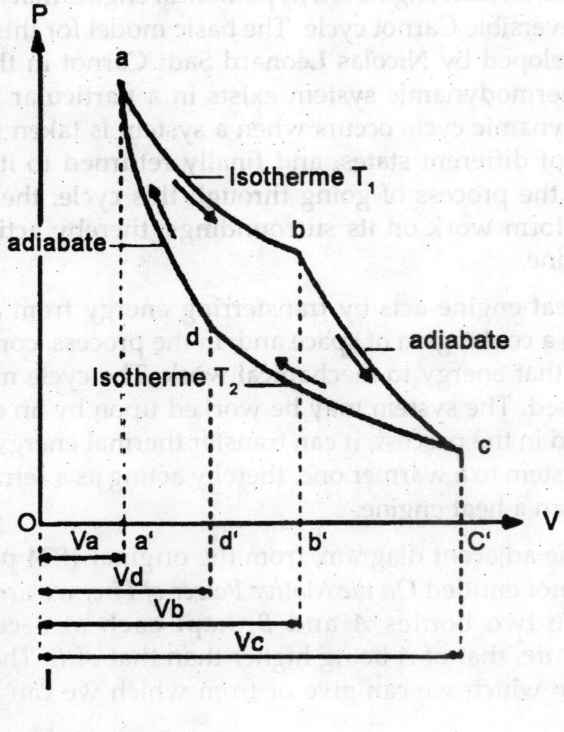

Carnot Engine

Engine using a Carnot cycle of operations. This type of engine can serve as the model for a thermic or refrigerator engine. An "animation" facilitates a better understanding of the cycle:

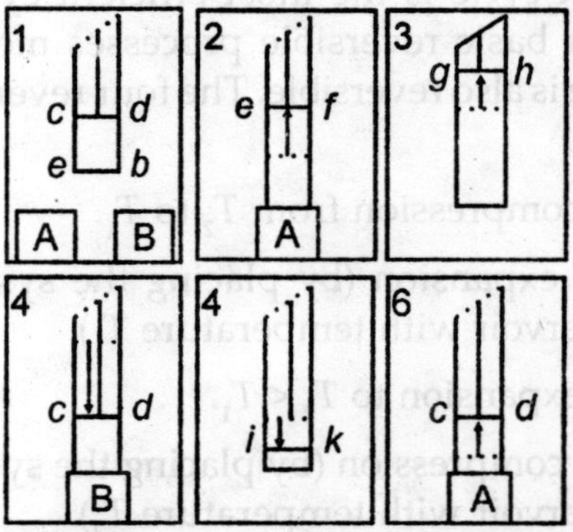

A Carnot heat engine is a hypothetical engine that operates on the reversible Carnot cycle. The basic model for this engine was developed by Nicolas Léonard Sadi Carnot in the 1824. Every thermodynamic system exists in a particular state. A thermodynamic cycle occurs when a system is taken through a series of different states, and finally returned to its initial state. In the process of going through this cycle, the system may perform work on its surroundings, thereby acting as a heat engine.

A heat engine acts by transferring energy from a warm region to a cool region of space and, in the process, converting some of that energy to mechanical work. The cycle may also be reversed. The system may be worked upon by an external force, and in the process, it can transfer thermal energy from a cooler system to a warmer one, thereby acting as a refrigerator rather than a heat engine.

In the adjacent diagram, from the original 1824 paper by Sadi Carnot entitled *On the Motive Power of Fire,* we are told to "imagine two bodies *A* and *B*, kept each at a constant temperature, that of *A* being higher than that of *B*. These two bodies, to which we can give or from which we can remove the heat without causing their temperatures to vary, exercise the functions of two unlimited reservoirs of caloric. We shall call the first the furnace and the second the refrigerator." Carnot then explains how we can obtain motive power, i.e. "work", by carrying a certain quantity of heat from the body *A* to the body *B*.

Carnot's Theorem

During one part of the cycle performed in an engine, some heat is absorbed from a hot reservoir. During another part, a smaller amount of heat is rejected to a cooler reservoir. The engine is therefore said to operate between these two reservoirs. It is a fact of experience that some heat is always rejected to the cooler reservoir. Because of this, the efficiency of an actual engine is never 100per cent. The processes that make up this cycle are either isothermal or adiabatic.

The Carnot Machine

We consider the standard Carnot-cycle machine, which can be thought of as having a piston moving within a cylinder, and having the following characteristics:

- A perfect seal, so that no atoms escape from the working fluid as the piston moves to expand or compress it.
- Perfect lubrication, so that there is no friction.
- An ideal-gas for the working fluid.
- Perfect thermal connection at any time either to one or to none of two reservoirs, which are at two different temperatures, with perfect thermal insulation isolating it from all other heat transfers.
- The piston moves back and forth repeatedly, in a cycle of alternating "isother-mal" and "adiabatic" expansions and compressions, according to the PV diagram shown:

By definition, the isothermal segments (AB and CD) occur when there is perfect thermal contact between the working fluid and one of the reservoirs, so that whatever heat is needed to maintain constant temperature will flow into or out of the working fluid, from or to the reservoir.

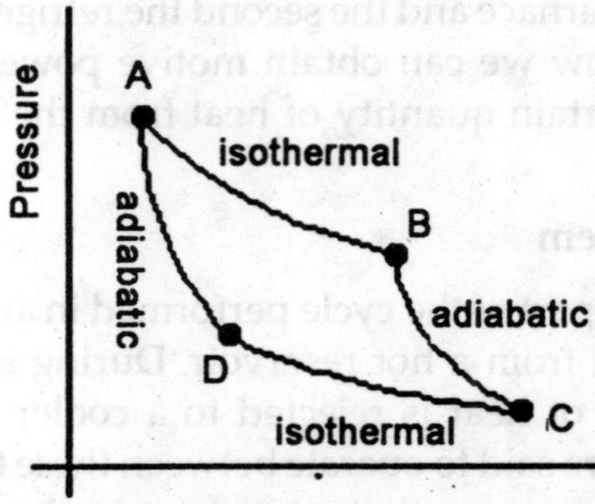

By definition, the adiabatic segments (BC and DA) occur when there is perfect thermal insulation between the working fluid and the rest of the universe, including both reservoirs, thereby preventing the flow of any heat into or out of the working fluid.

The isothermal curves (but not the adiabatic curves) are hyperbolas, according to PV = nRT. The enclosed area (and therefore the mechanical work done) will depend on the two temperatures ("height") and on the amount of heat transferred, which depends in turn on the extent of the isothermal compression or expansion ("width"), during which heat must be transferred to maintain the constant temperature.

We will denote the heat transferred to or from the high-temperature reservoir (during the transition between points A and B) as Qh.

We will denote the heat transferred to or from the low-temperature reservoir (during the transition between points C and D) as Qc.

Engine, Heat Pump or Refrigerator

If a Carnot machine cycles around the path clockwise, a high-temperature isothermal expansion from A to B, an adiabatic expansion cooling down from B to C, a low-temperature isothermal compression from C to D, and finally an adiabatic compression warming up from D to A, it functions as a heat engine, removing energy from the high-temperature reservoir as heat, transforming a portion of that energy to useful mechanical work (the enclosed area) done on the external world, and ejecting the remainder of the energy as waste heat to the low-temperature reservoir.

If a Carnot machine is driven (by an external agency, such as a motor) around the cycle counter clockwise, an adiabatic expansion cooling down from A to D, a low-temperature isothermal expansion from D to C, an adiabatic compression warming up from C to B, and finally a high temperature isothermal compression from B to A, then it functions as either a refrigerator or a heat pump, depending on whether removing heat from the low-temperature reservoir or adding heat to the high-temperature reservoir is of primary interest. The mechanical energy required to force the machine around the cycle is the work done on the machine, the area enclosed.

Efficiency

For a heat engine, the efficiency is the ratio of useful work performed to the heat energy consumed from the high-temperature reservoir:

$$Eff = \frac{W}{Q_h}$$

$$= \frac{Q_h - Q_c}{Q_h}$$

This ratio is the interesting one because you pay for the fuel to obtain Qh, in order to get the benefit of the work done, W. For a Carnot engine, this is entirely determined by the temperatures of the hot and cold reservoirs:

$$Eff = \frac{T_h - T_c}{T_h}$$

$$= 1 - \frac{T_c}{T_h}$$

This temperature dependence is a direct consequence of the second law of thermodynamics and the fact that all heat transfers occur during isothermal expansion and contraction, with no temperature difference between the heat reservoir and the working fluid, so that the entropy gained by one exactly matches the entropy lost by the other, with no net change in entropy for the system as a whole. This condition is of course an ideal one, and cannot be met in practice by any real machine. Thus, the Carnot efficiency is the best possible even theoretically; all real machines will be strictly worse than this.

For a Carnot machine functioning as a refrigerator, the "effectiveness" is the ratio of the energy removed from the low-temperature reservoir to the work required to force the machine around its cycle (the energy consumed and paid for):

$$Eff = \frac{Q_c}{W}$$

$$= \frac{Q_c}{Q_h - Q_c}$$

The effectiveness will be greater than 1 only if the absolute temperature of the cold reservoir is warmer than half that of the hot reservoir. We can see that refrigeration to extremely cold temperatures is very difficult.

For a Carnot machine functioning as a heat pump, the "effectiveness" is the ratio of the energy delivered to the high-temperature reservoir to the work required to force the machine around its cycle (the energy consumed and paid for):

$$Eff = \frac{Q_h}{W}$$

$$= \frac{Q_h}{Q_h - Q_c}$$

This effectiveness is also known as the coefficient of performance ("COP"). For heat pumps, the effectiveness is always greater than 1. Electrically powered heat pumps can make economic sense only if the effectiveness of the heat pump times the efficiency of the electrical generation and transmission process exceeds 1. Otherwise, only part of the fuel burned to produce the electricity would have to be burned to provide the heat needed. (Modern natural gas furnaces can easily transfer more than 95per cent of the combustion heat to the heated space.) As the temperature of the cold reservoir (the outside temperature) declines, the COP of the heat pump decreases toward 1. Because large electrical generators produce about one-third as much electical energy as the heat value of the fuel they consume, as soon as the COP is less than about 3, it would be cheaper to burn the original fuel directly for the heat, rather than generate electricity to operate a heat pump. This limits the geographical regions where heat pumps make economic sense.

Real Process Cycle Compared to Carnot Cycle

The most important aspect of the second law for our practical purposes is the determination of maximum possible efficiencies obtained from a power system. Actual efficiencies will always be less than this maximum. The losses (friction, for example) in the system and the fact that systems are not truly reversible preclude us from obtaining the maximum possible efficiency.

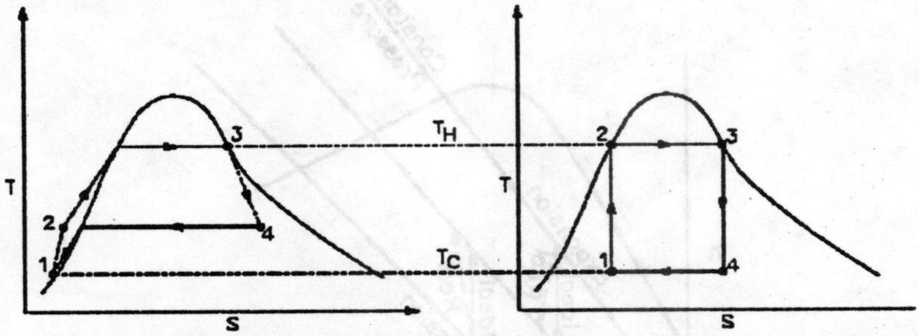

Any ideal thermodynamic process can be drawn as a path on a property diagram, such as a T-s or an h-s diagram. A real process that approximates the ideal process can also be represented on the same diagrams (usually with the use of dashed lines). In an ideal process involving either a reversible expansion or a reversible compression, the entropy will be constant. These isentropic processes will be represented by vertical lines on either T-s or h-s diagrams, since entropy is on the horizontal axis and its value does not change. A real expansion or compression process

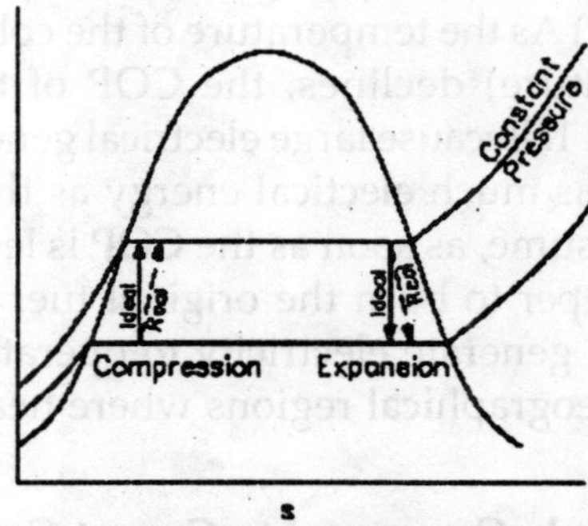

Fig. (a) Expansion and Compression Processes on T-s Diagram

operating between the same pressures as the ideal process will look much the same, but the dashed lines representing the real process will slant slightly towards the right since the entropy will increase from the start to the end of the process. Figures shown (a and b) below shows ideal and real expansion and compression processes on T-s and h-s diagrams.

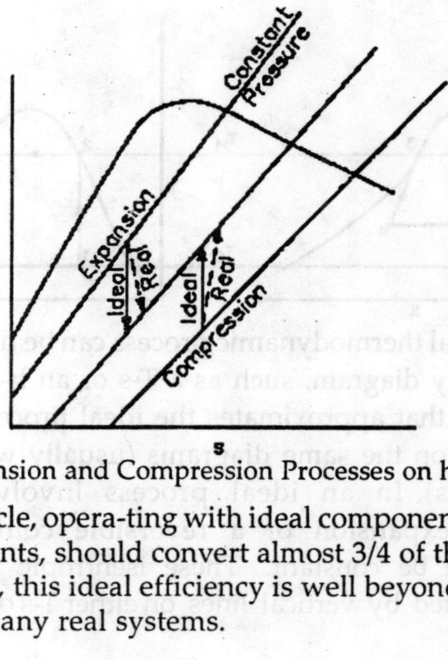

Fig. (b) Expansion and Compression Processes on h-s Diagram

Carnot cycle, opera-ting with ideal components under real world constraints, should convert almost 3/4 of the input heat into work. But, this ideal efficiency is well beyond the present capabilities of any real systems.

Early thermo-dyna-mic developments were centered around impro-ving the perfor-mance of contem-porary steam engines. It was desirable to construct a cycle that was as close to being reversible as possible and would better lend itself to the characteristics of steam and process control than the Carnot cycle did. Towards this end, the Rankine cycle was developed.

Rankine Cycle

The main feature of the Rankine cycle, is that it confines the isentropic compression process to the liquid phase only. This minimizes the amount of work required to attain operating pressures and avoids the mechanical problems associated with pumping a two-phase mixture. The Rankine cycle is a thermodynamic cycle. Like other thermodynamic cycles, the maximum efficiency of the Rankine cycle is given by calculating the maximum efficiency of the Carnot cycle. It is named after William John Macquorn Rankine.

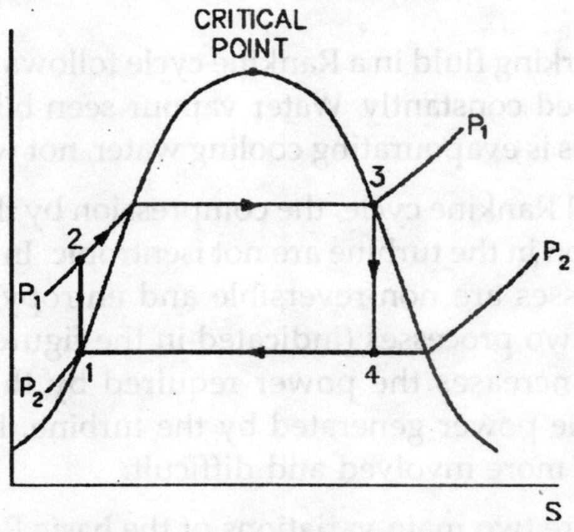

There are four processes in the Rankine cycle, each changing the state of the working fluid. These states are identified by number in the diagram above.

- *Process 1-2*: First, the working fluid is pumped (ideally isentropically) from low to high pressure by a pump. Pumping requires a power input (for example mechanical or electrical).

- *Process 2-3*: The high pressure liquid enters a boiler where it is heated at constant pressure by an external heat source to become a superheated vapour. Common heat sources for power plant systems are coal, natural gas, or nuclear power.
- *Process 3-4*: The superheated vapour expands through a turbine to generate power output. Ideally, this expansion is isentropic. This decreases the temperature and pressure of the vapour.
- *Process 4-1*: The vapour then enters a condenser where it is cooled to become a saturated liquid. This liquid then re-enters the pump and the cycle repeats.

Rankine cycles describe the operation of steam heat engines commonly found in power generation plants. In such vapour power plants, power is generated by alternately vapourizing and condensing a working fluid (in many cases water, although refrigerants such as ammonia may also be used).

The working fluid in a Rankine cycle follows a closed loop and is re-used constantly. Water vapour seen billowing from power plants is evapourating cooling water, not working fluid.

In a real Rankine cycle, the compression by the pump and the expansion in the turbine are not isentropic. In other words, these processes are non-reversible and entropy is increased during the two processes (indicated in the figure as ΔS). This somewhat increases the power required by the pump and decreases the power generated by the turbine. It also makes calculations more involved and difficult.

There are two main variations of the basic Rankine cycle:

Rankine Cycle with Reheat

In this variation, two turbines work in series. The first accepts vapour from the boiler at high pressure. After the vapour has passed through the first turbine, it re-enters the boiler and is reheated before passing through a second, lower pressure turbine. Among other advantages, this prevents the vapour from condensing during its expansion which can seriously damage the turbine blades.

Regenerative Rankine Cycle

The regenerative Rankine cycle is so named because after emerging from the condenser the working fluid is heated by steam tapped from the hot portion of the cycle. This increases the average temperature of heat addition which in turn increases the thermodynamic efficiency of the cycle.

Steam Cycle

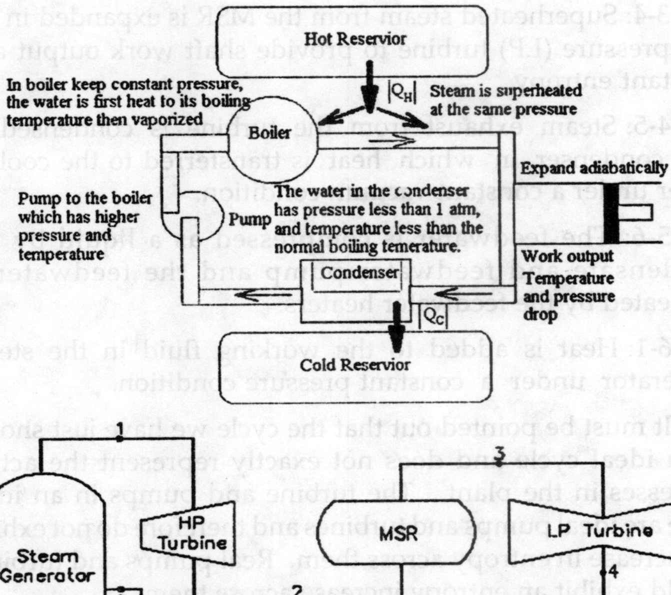

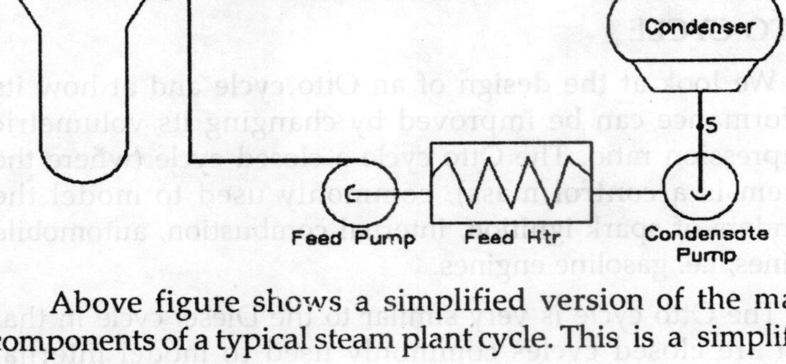

Above figure shows a simplified version of the major components of a typical steam plant cycle. This is a simplified version and does not contain the exact detail that may be found at most power plants.

The following are the processes that comprise the cycle:

1-2: Saturated steam from the steam generator is expanded in the high pressure (HP) turbine to provide shaft work output at a constant entropy.

2-3: The moist steam from the exit of the HP turbine is dried and superheated in the moisture separator reheater (MSR).

3-4: Superheated steam from the MSR is expanded in the low pressure (LP) turbine to provide shaft work output at a constant entropy.

4-5: Steam exhaust from the turbine is condensed in the condenser in which heat is transferred to the cooling water under a constant vacuum condition.

5-6: The feedwater is compressed as a liquid by the condensate and feedwater pump and the feedwater is preheated by the feedwater heaters.

6-1: Heat is added to the working fluid in the steam generator under a constant pressure condition.

It must be pointed out that the cycle we have just shown is an ideal cycle and does not exactly represent the actual processes in the plant. The turbine and pumps in an ideal cycle are ideal pumps and turbines and therefore do not exhibit an increase in entropy across them. Real pumps and turbines would exhibit an entropy increase across them.

OTTO CYCLE

We look at the design of an Otto cycle and at how its performance can be improved by changing its volumetric compression ratio. The Otto cycle a closed cycle (where the system is a control mass), commonly used to model the cylinders of spark-ignition, internal combustion, automobile engines, i.e. gasoline engines.

The Otto cycle is very similar to the Diesel cycle in that both are closed cycles commonly used to model internal combustion engines. The difference between them is that the Otto cycle is a spark-ignition cycle instead of a compression-ignition cycle like the Diesel cycle. Spark-ignition cycles are designed to use fuels that require a spark to begin combustion.

Stages

Otto Cycles have four stages: expansion, cooling, compression, and combustion.

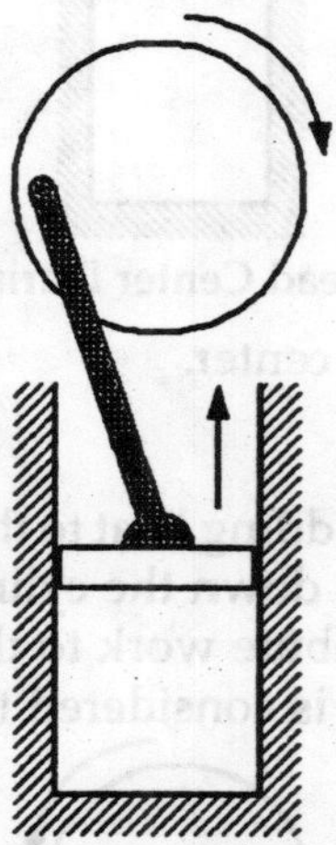

Fig. Moving up During Expansion

Expansion

In the Otto cycle, fuel is burned to heat compressed air and the hot gas expands forcing the piston to travel up in the cylinder. It is in this phase that the cycle contributes its useful work, rotating the automobile's crankshaft. We make the ideal assumption that this stage in an ideal Otto cycle is isentropic.

Piston: moving from bottom dead center to top dead center.

Cooling

Next, the expanded air is cooled down to ambient conditions. In an actual automobile engine, this corresponds to exhausting the air from the engine to the environment and replacing it with fresh air. Since this happens when the piston is at the top dead center position in the cycle and is not moving, we say this process is isochoric (no change in volume).

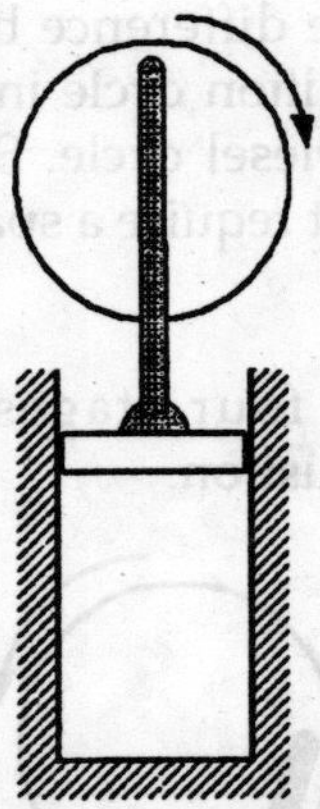

Fig. Top Dead Center During Cooling

Piston: at top dead center.

Compression

In preparation for adding heat to the air, we next compress it by moving the piston down the cylinder. It is in this part of the cycle that we contribute work to the air. In the ideal Otto cycle, this compression is considered to be isentropic.

Fig. Moving Down During Compression

It is at this stage that we set the volumetric compression ratio, *r* which is the ratio of the volume of the working fluid before the compression process to its volume after. It will turn out that the efficiency of the Otto cycle (assuming that air is an ideal gas) can be described entirely in terms of this ratio.

Piston: moving from top dead center to bottom dead center.

Combustion

Next, heat is added to the air by fuel combustion when the piston is at it bottom dead center position. Combustion is not initiated until a spark (from a spark plug, for instance) is generated in the cylinder. Because the piston is essentially immobile during this part of the cycle, we say that the heat addition is isochoric, like the cooling process.

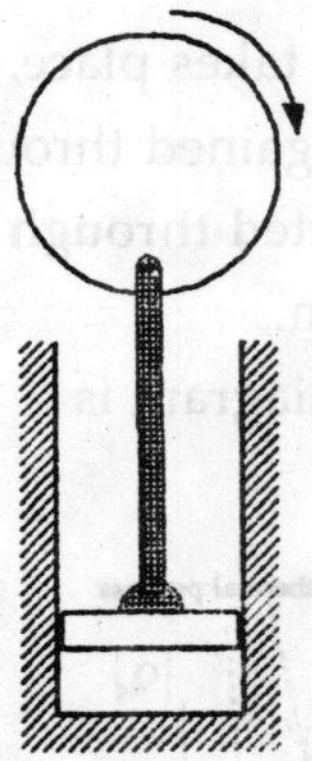

Fig. Bottom Dead Center During Combustion

Piston: at bottom dead center.

P-V Diagram

The P-V diagram for an Otto cycle is shown below.

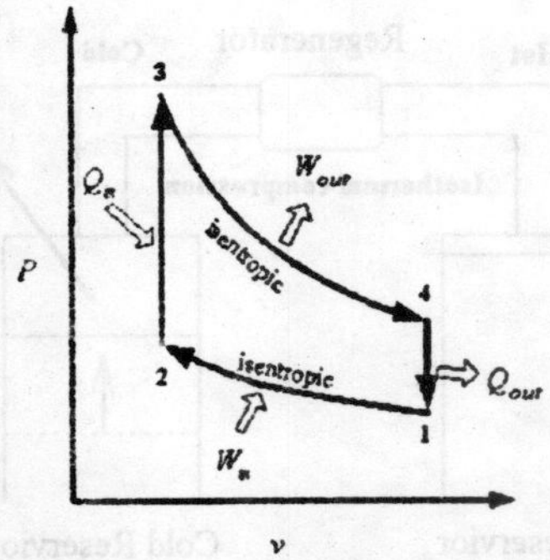

Fig. Otto Cycle P-V diagram

THE STIRLING ENGINE

In 1816, before the science of thermodynamics had been begun, a minister of the Church of Scotland named Robert Stirling designed a hot-air engine which can convert some of the heat from burning fuel into work. Advantage of this engine is low exhaust emission and high efficiencies. Disadvantage of this engine is high manufacturing costs. Idealized Stirling Engine are shown in the following, which is based on some assumptions, and they are:

1. The gas is ideal,
2. No leakage of gas takes place,
3. No heat is lost or gained through cylinder walls,
4. No heat is conducted through the regenerator,
5. There is no friction.

The Stirling Cycle diagram is

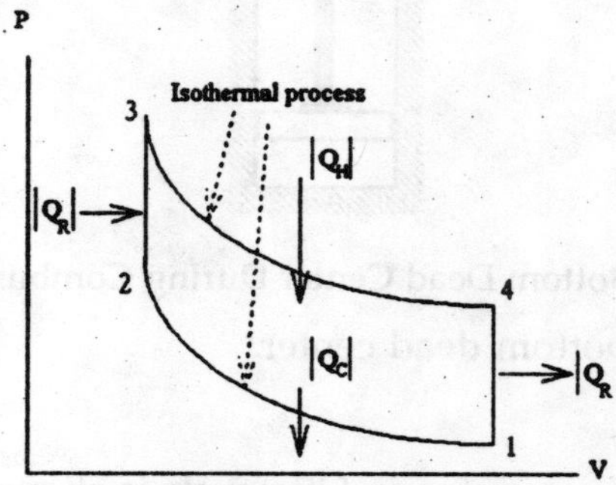

Form 1 → to 2

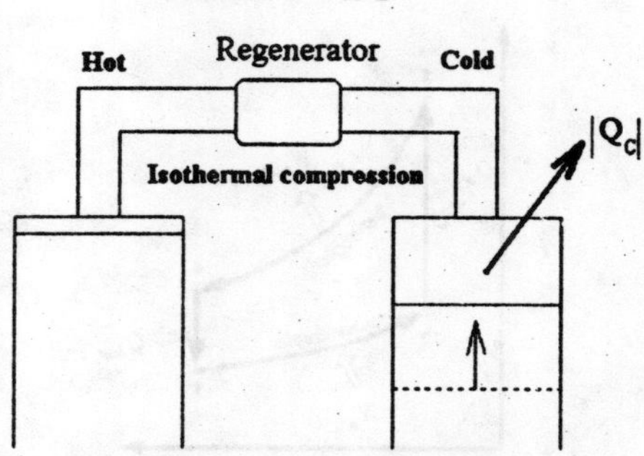

12 → It is an isothermal process, the piston in contact with cold reservoir is compressed isothermally, hence heat $|QC|$ has been rejected, and (isothermal compression → $dU = 0$, W is positive and QC is negative) the heat rejected is

$$Q_C = \int_{V_1}^{V_2} PdV = nR\theta_C \ln \frac{V_2}{V_1}$$

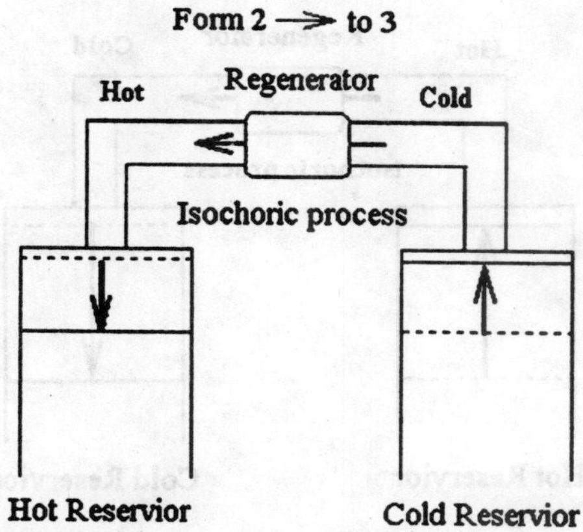

2 → 3 It is an isochoric process, the left piston moves down while the right piston moves up. The volume of system is kept constant, thus no work has been done by the system, but heat QR has been input to the system by the regenerator which causes temperature to raise to θH.

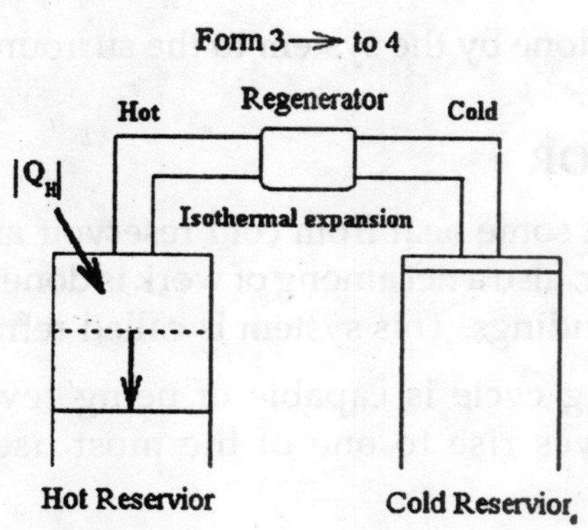

$3 \rightarrow 4$ It is an isothermal expansion process, the left piston in contact with hot reservoir expanded isothermally at temperature θH. Therefore

$$Q_H = \int_{V_3}^{V_4} PdV = nR\theta_H \ln \frac{V_4}{V_3}$$

Form 4 ⟶ to 1

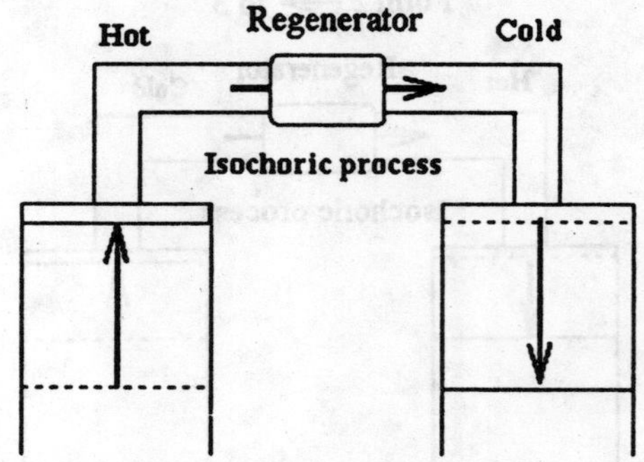

$4 \rightarrow 1$ It is an isochoric process which is a reversed process of $2 \rightarrow 3$, but from θH to θC. The efficiencies of Stirling engine is

$$n = 1 - \frac{|Q_C|}{Q_H} = 1 - \frac{\theta_C}{\theta_H}$$

and work done by the system to the surroundings is |W| = |QH| – |QC|.

REFRIGERATOR

Absorption some heat from cold reservoir and reject it to the hot reservoir, also a net among of work is done to the system from its surroundings. This system is called refrigerant.

The Stirling cycle is capable of being reversed, when reversed, it gives rise to one of the most useful types of refrigerator.

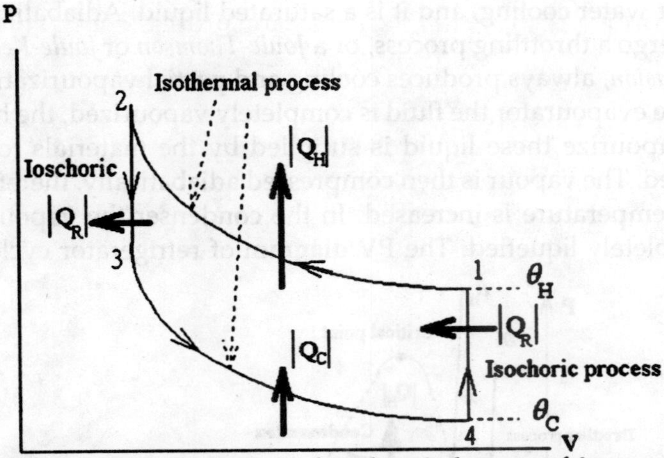

The schematic diagram for electric home refrigerators is shown below, a constant mass of refrigerant is stored in the liquid storage which have the same temperature and pressure as it in the condensor. The refrigerant go through the throttling valve, through the evapourator, into the compressor and finially back to the condensor.

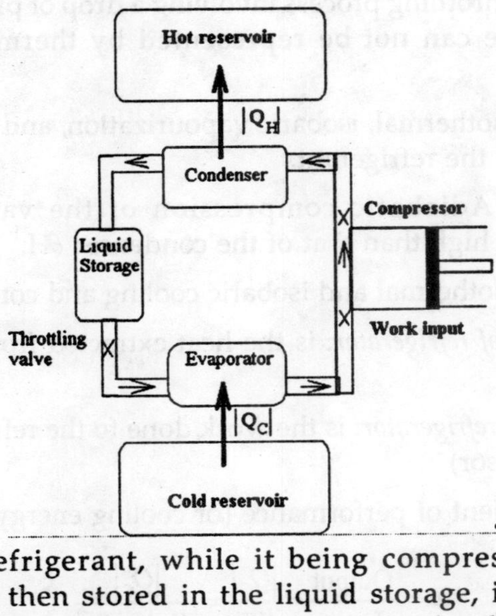

The refrigerant, while it being compressed in the condenser then stored in the liquid storage, is at a high pressure and at as low a temperature as can be obtained with air or water cooling, and it is a saturated liquid. Adiabatically undergo a throttling process, or a *Joule-Thomson* or *Joule-Kelvin expansion*, always produces cooling and partial vapourization. In the evapourator the fluid is completely vapourized, the heat to vapourize these liquid is supplied by the materials to be cooled. The vapour is then compressed adiabatically, therefore the temperature is increased. In the condenser the vapour is completely liquefied. The PV diagram of refrigerator cycle is

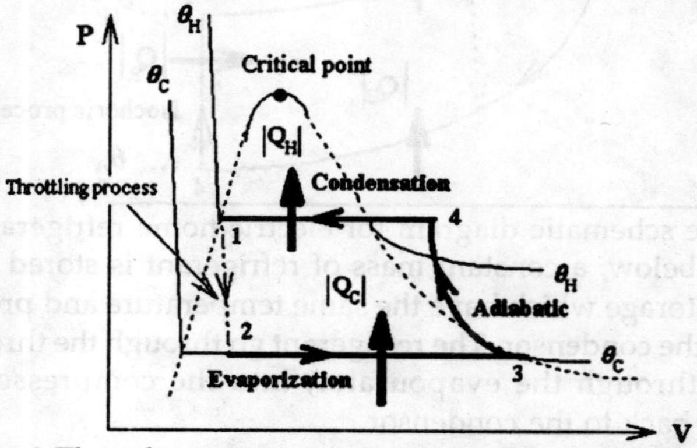

1 → 2 Throttling process involving a drop of pressure and temperature can not be represented by thermodynamic coordinates.

2 → 3 Isothermal, isobaric vapourization, and heat *QC* is absorbed by the refrigerant.

3 → 4 Adiabatic compression of the vapour to a temperature high than that of the condenser *θ*H.

4 → 1 Isothermal and isobaric cooling and condensation.

Output of refrigerator: is the heat extracted from the cold reservoir.

Input of refrigerator: is the work done to the refrigerant by the compressor)

Coefficient of performance (or cooling energy ratio) ω

$$\omega = \frac{\text{Output}}{\text{Input}} = \frac{|Q_C|}{|W|} = \frac{|Q_C|}{|Q_H| - |Q_C|}$$

ω for commerical build refrigerator is about 2 to 7. For the case of $\omega = 5$, that is

$$\omega = \frac{|Q_C|}{W} = 5 \Rightarrow \frac{|Q_H| - |W|}{|W|} = 5$$

which gives

$$\frac{|Q_H|}{W} = 6,$$

this means the heat liberated at the higher temperature is equal to six times the work done. In this case, if the work is supplied by an electric motor, for every joule of electrical energy supplied, 6 J of heat will liberated, it is a very good ideal to build a heater to warm the house by refrigerating the outdoors. (In general, use resister as heater, which 1 J of electrical energy can only give 1 Joule of heat.) This heater idea is first point out by Lord Kelvin, but the first heater to heat a house is built by Haldane.

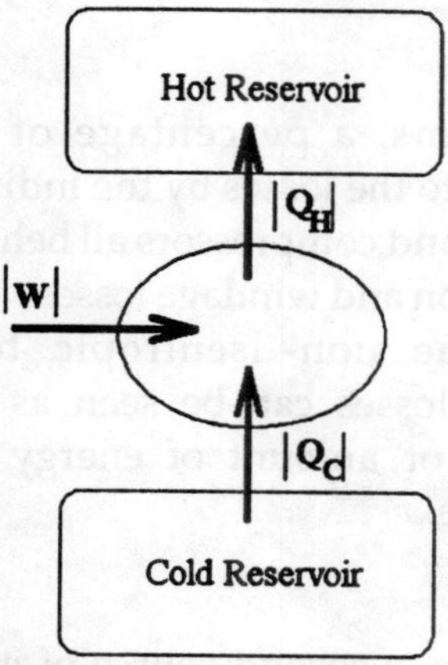

ERICSSON CYCLE

The Ericsson Cycle is named after inventor John Ericsson. The Ericsson cycle is very similar to what we now call the "Brayton Cycle".

The main difference is that the all Ericsson engines are externally heated and usually have a recuperator or regenerator between the compressor and the expander. The Ericsson cycle is often compared to the Stirling Cycle because of its external combustion capabilities, use of a regenerator and the high potential efficiency.

Barber was the first to propose a similar engine in 1791 but the Barber engine lacked a regenerator/ recuperator. Ericsson invented and patented his first engine in 1833 (number 6409/1833 British). This was 18 years before Joule and 43 years before Brayton. Brayton engines were for the most part an internal combustion version of the Ericsson Cycle which was a considerable improvement to external heating. The "Brayton Cycle" is now known as the gas turbine cycle, which differs from the original "Brayton Cycle" in the use of a turbine compressor and expander. The gas turbine cycle is used for all modern gas turbine and turbojet engines. Ericsson eventually abandoned the open cycle in favor of the traditional closed Stirling cycle.

CAUSES OF INEFFICIENCY

Components

In real systems, a percentage of the overall cycle inefficiency is due to the losses by the individual components. Turbines, pumps, and compressors all behave non-ideally due to heat losses, friction and windage losses. All of these losses contribute to the non- isentropic behaviour of real equipment. These losses can be seen as an increase in the system's entropy or amount of energy that is unavailable for use by the cycle.

Cycles

In real systems, a second source of inefficiencies is from the compromises made due to cost and other factors in the design and operation of the cycle. Examples of these types of losses are: In a large power generating station the condensers are designed to subcool the liquid by 8-10°F. This subcooling allows the condensate pumps to pump the water forward without cavitation. But, each degree of subcooling is energy that must be put back by reheating the water, and this heat (energy) does no useful work and therefore increases the inefficiency of the cycle. Another example of a loss due to a system's design is heat loss to the environment, i.e. thin or poor insulation. Again this is energy lost to the system and therefore unavailable to do work. Friction is another real world loss, both resistance to fluid flow and mechanical friction in machines. All of these contribute to the system's inefficiency. Efficiency will be decreased by: Presence of friction, Heat losses, Cycle inefficiencies.

Chapter 10

Work Transfer

Work is defined as force acting upon an object to cause a displacement. There are three key words in this definition - force, displacement, and cause. In order for a force to qualify as having done *work* on an object, there must be a displacement and the force must *cause* the displacement. There are several good examples of work which can be observed in everyday life - a horse pulling a plow through the fields, a father pushing a grocery cart down the aisle of a grocery store, a freshman lifting a backpack full of books upon her shoulder, a weightlifter lifting a barbell above her head, an Olympian launching the shot-put, etc. In each case described here there is a force exerted upon an object to cause that object to be displaced.

Mathematically, work can be expressed by the following equation.

$$W = F \times d \times \cos\theta$$

where F = force, d = displacement, and the angle (theta) is defined as the angle between the force and the displacement vector. Perhaps the most difficult aspect of the above equation is the angle "theta.". The angle measure is defined as the angle between the force and the displacement. To gather an idea of its meaning, consider the following three casses.

- *A*: A force acts rightward upon an object as it is displaced rightward. In such an instance, the force vector and the displacement vector are in the same direction. Thus, the angle between F and d is 0 degrees.

- *B*: A force acts leftward upon an object which is displaced rightward. In such an instance, the force vector and the displacement vector are in the opposite direction. Thus, the angle between F and d is 180 degrees.
- C: A force acts upward upon an object as it is displaced rightward. In such an instance, the force vector and the displacement vector are at right angles to each other. Thus, the angle between F and d is 90 degrees.

Let's consider case C above in more detail. case C involves a situation similar to the waiter who carried a tray full of meals above his head by one arm straight across the room at constant speed. The force supplied by the waiter on the tray is an upward force and the displacement of the tray is a horizontal displacement. As such, the angle between the force and the displacement is 90 degrees. If the work done by the waiter on the tray were to be calculated, then the results would be 0. Regardless of the magnitude of the force and displacement, F × d ×cosine 90 degrees is 0 (since the cosine of 90 degrees is 0). A vertical force can never cause a horizontal displacement; thus, a vertical force does not do work on a horizontally displaced object.

It can be accurately noted that the waiter's hand did push forward on the tray for a brief period of time to accelerate it from rest to a final walking speed. But once *up to speed*, the tray wil! stay in its straight-line motion at a constant speed without a forward force. And if the only force exerted upon the tray during the constant speed stage of its motion is upward, then no work is done upon the tray. Again, a vertical force does not do work on a horizontally displaced object.

The equation for work lists three variables - each variable is associated with one of the three key words mentioned in the definition of work (force, displacement, and cause). The angle theta in the equation is associated with the amount of force which causes a displacement. When a force is exerted on an object at an angle to the horizontal, only a part of the force contributes to (or causes) a horizontal displacement.

The Meaning of Theta

When determining the measure of the angle in the work equation, it is important to recognize that the angle has a precise definition - it is the angle between the force and the displacement vector. For instance, consider the activity performed in the given figure. A force was applied to a cart to pull it up an incline at constant speed. Several incline

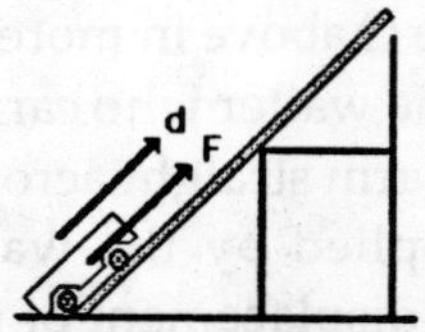

Whenever F and d are in the same direction, $\theta = 0$ degrees.

angles were used; yet, the force was always applied parallel to the incline. The displacement of the cart was also parallel to the incline. Since F and d were in the same direction, the angle was 0 degrees. The angle in the equation is defined as the angle between the force and the displacement vector.

Units

In the case of work (and also energy), the standard metric unit is the Joule (abbreviated "J"). One Joule is equivalent to one Newton of force causing a displacement of one meter. In other words,

The Joule is the unit of work.

1 Joule = 1 Newton × 1 meter

1 J = 1 Nm

In fact, any unit of force times any unit of displacement is equivalent to a unit of work. Some nonstandard units for work are shown below. Notice that when analysed, each set of units is equivalent to a force unit times a displacement unit.

Non-SI units of work include the erg, the foot-pound, the foot-poundal, and the liter-atmosphere.

FLUID

A fluid is any substance that conforms to the shape of its container. It may be either a liquid or a gas. A fluid is anything that would spill or float away if it weren't in a container (unless it's big enough to be held together by gravity like a star). If you can stir it up with a spoon or blow it through a straw, it's a fluid. Water is a fluid and so is air. In fact, all liquids and gases are fluids. In space and inside stars there's also another kind of fluid called a plasma.

The molecules in a solid are stuck together, but in a fluid they're free to move past each other. So if you had very small hands you could push one molecule of a fluid one way and another molecule the other way and off they would go in the direction you pushed them.

Compressibility of Fluids

All fluids are compressible. However, under some range of conditions, it is often possible to make the approximation that a fluid is incompressible. Water, for example, only changes its volume very slightly under extreme pressure.

Gases are much more compressible. The compressibility of air, for example, is part of our common experience. By blocking off a bicycle pump and pushing down on the handle, we can easily decrease the volume of the air by 50per cent, so that its density increases by a factor of two (the mass of air is constant). If we assume that the temperature remains constant (somehow), we know from the perfect gas law,

$$p = \rho RT$$

that the pressure must also increase by a factor of two. If the initial air was at atmospheric pressure, the pressure will rise by one atmosphere (14.696 psi, or 101,325 Pa). If we assume a pump diameter of 1.25 inch, then a force of 18.1 lbf will be required. This is not a large force, so that raising the pressure by two atmospheres above ambient pressure is easily done.

We call pressures relative to zero pressure absolute pressure. This is the pressure that appears in the perfect gas law. Pressures measured relative to atmospheric pressure are called gauge pressures. The pressure measured by the most common types of pressure instruments is a gauge pressure since these instruments indicate the pressure relative to atmospheric pressure. A tire gauge, for instance, measures the pressure in a tire over and above the local atmospheric pressure. A vacuum gauge, in contrast, will measure the pressure below atmospheric: in common parlance a "vacuum" is any pressure lower than the ambient atmospheric pressure.

Even though gases are much more compressible than liquids (by perhaps a factor of 10^4), if the pressures are small the changes in density are also small. For example, if a 1per cent change in density is tolerated, then at constant temperature we can allow a 1per cent change in pressure. For air this corresponds to a change in altitude of about 85 meters, so that for changes in height of the order of tall buildings we can usually neglect the compressibility of air. The compressibility of gases is also often discussed in terms of the parameter called the Mach number, which is the ratio of the fluid velocity to the speed of sound (the speed of sound for air at 20ºC is 343 m/s = 1,126 ft/s = 768 mph). When a fluid increases its speed, there is a corresponding decrease in pressure. If the fluid velocity is small relative to the speed of sound (that is, the flow has a low Mach number), then the change in pressure when a fluid increases its velocity from zero to V is given by 1/2 the density times the velocity squared (in the absence of friction). This quantity is called the dynamic pressure:

$$\text{dynamic pressure} = \frac{1}{2}\rho V^2$$

If we again use a 1per cent level as a tolerable change in density, then if the temperature remains constant the allowed change in pressure is also equal to 1per cent (the process is more likely to be isentropic, but that does not change the argument very much). This requires a dynamic pressure less than 1per cent of the ambient pressure, so that at sea level where the density of air is about 1.2 kg/m^3 we are limited to velocities less than 40 m/s (132 ft/s or 90 mph), which corresponds to a Mach number of about 0.12. When we consider the flow of air over bicycles, we can always assume that the fluid has a constant density, and that the Mach number is not important.

CONSTANT PRESSURE PROCESS

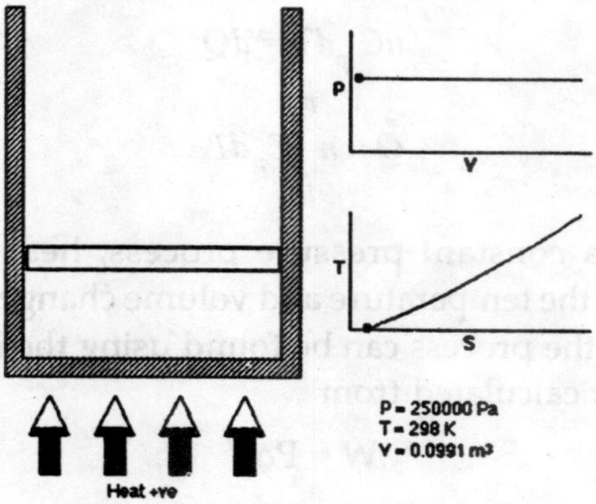

Applying the first law of thermodynamics to the process

$$dU = dQ - dW$$

Replacing dW with the reversible work

$$dU = dQ - PdV$$

The volume will change as the gas is heated at constant pressure. To make calculations more straight forward, use ENTHALPY, H

$$H = U + PV$$

$$dH = dU + PdV + VdP$$

rearrange for dU

$$dU = dH - PdV - VdP$$

and substitute into the first law

$$dH - PdV - VdP = dQ - PdV$$

the PdV terms cancel out and since pressure is constant dP = 0, so that

$$dH = dQ$$

The definition of the specific heat at constant pressure

$$C_P = \frac{dH}{dT}$$

is used to replace dH in the first law

$$nC_p dT = dQ$$

$$Q = n\int_{T_1}^{T_2} C_p dT$$

During a constant pressure process, heat is added or removed and the temperature and volume change. The volume at the end of the process can be found using the ideal gas law and the work calculated from

$$W = P\Delta V$$

Entropy Change

Starting with the first law expression for the process

$$dH = dQ$$

and replacing dH from the definition of specific heat at constant pressure and Dq from the definition of entropy

$$nC_p dT = TdS$$

$$\Delta S = n\int_{T_1}^{T_2} \frac{C_p}{T} dT$$

CONSTANT VOLUME PROCESS

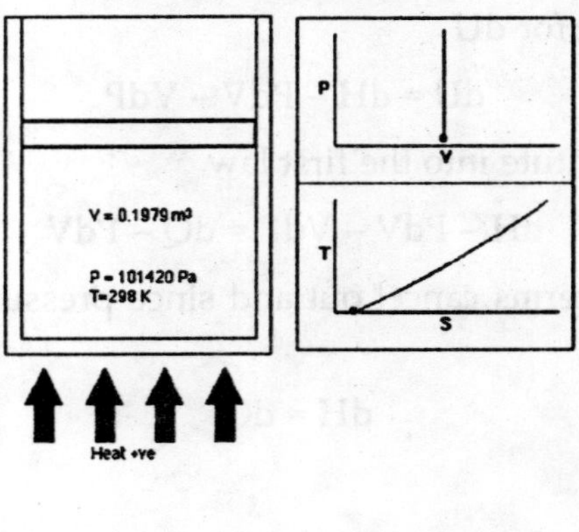

Applying the first law of thermodynamics to the process

$$dU = dQ - dW$$

Replacing dW with the reversible work

$$dU = dQ - PdV$$

since the volume is constant dV = 0 and

$$dU = dQ$$

using the definition of the specific heat at constant volume

$$C_v = \frac{dU}{dT}$$

to replace dU in the first law

$$dQ = nC_v dT$$

$$Q = n\int_{T_1}^{T_2} C_V dT$$

For a constant volume process, the addition or removal of heat will lead to a change in the temperature and pressure of the gas, as shown on the two graphs above

Entropy Change

To find the Entropy change, start with the expression derived from the first law

$$\mathbf{dU = dQ}$$

and replace **dU** using the definition of specific heat at constant volume and **dQ** using the definition of entropy

$$nC_V dT = TdS$$

$$\Delta S = n\int_{T_1}^{T_2} \frac{C_V}{T} dT$$

Effects of Pressure Changes on Fluid Properties

The predominant effect of an increase in pressure in a compressible fluid, such as a gas, is an increase in the density of the fluid. An increase in the pressure of an incompressible fluid will not have a significant effect on the density. For example, increasing the pressure of 100 °F water from 15 psia to 15,000 psia will only increase the density by approximately 6per cent. Therefore, in engineering calculations, it is assumed that incompressible fluids' density remain constant.

Effects of Temperature Changes on Fluid Properties

An increase in temperature will tend to decrease the density of any fluid. If the fluid is confined in a container of fixed volume, the effect of a temperature change will depend on whether the fluid is compressible.

If the fluid is a gas, it will respond to a temperature change in a manner predicted by the ideal gas laws. A 5per cent increase in absolute temperature will result in a 5per cent increase in the absolute pressure.

If the fluid is an incompressible liquid in a closed container, an increase in the temperature will have a tremendously greater and potentially catastrophic effect.

As the fluid temperature increases, it tries to expand, but expansion is prevented by the walls of the container. Because the fluid is incompressible, this results in a tremendous increase in pressure for a relatively minor temperature change. The change in specific volume for a given change in temperature is not the same at various beginning temperatures.

Resultant pressure changes will vary. A useful thumb rule for water is that pressure in a water-solid system will increase about 100 psi for every 1 °F increase in temperature.

NON FLOW ENERGY EQUATION

Let a system boundary enclose only the water that is inside the reservoir. The underlying principle is that the net rate at which energy is transferred and transported into the system equals the rate of change of the energy of the system.

Internal energy is transported into and out of the system with the water flows and is lost from the system as heat transfer to the surroundings. When the relevant simplifying assumptions are made the situation is described by a first order differential equation in temperature. This can be integrated from the initial temperature of the water in the reservoir at time zero to give the temperature of the water at any future time. The flow rate of water into the system is steady. There is no flow out of the system until the overflow level is reached. Thereafter the flow rate out of the system is steady and equals the flow rate in. After a finite time the temperature of the water leaving will be within a specified tolerance of the eventual (infinite time) steady state value.

You may or may not agree with the simplifying assumptions mentioned here.

- The water can be regarded as an incompressible liquid.
- The specific heat of the water can be taken as constant.
- Kinetic energy of the water streams and the water in the reservoir can be neglected.
- Potential energy changes can also be neglected; they have no influence on the temperatures in this problem.
- Flow work terms (commonly dealt with by the use of the enthalpy property to include both flow work transfer and internal energy transport) and displacement work terms associated with the change in volume of the system boundary cancel out and do not affect the temperatures.
- A reservoir such as this may be subject to stratification and significant temperature variations may exist within it. For analysis purposes complete and continuous mixing of the water in the reservoir (without fluid friction) will be assumed.
- As a consequence of the ongoing mixing, the temperature of the water coming out of the reservoir is equal to the temperature of the water in the reservoir.

- From the mass balance point of view the loss of water from the surface can be neglected. However, the evapourative effect at the surface will be significant in energy terms (unless the air above the surface is saturated). It is common practice to deal with this as heat transfer between the water surface and the air.
- The rate of heat transfer from the water in the reservoir to the surroundings can be described and evaluated in terms of the temperature of the water, the temperature of the surroundings and the water level (if there is heat transfer through the side walls of the reservoir).

The non flow energy equation for the system can be written as follows:

$$\dot{m}_{\text{in}} u_{\text{in}} - \dot{m}_{\text{out}} u_{\text{out}} - \dot{Q}_{\text{out}} = \frac{dU}{dt}$$

$$= m\frac{du}{dt} + u\frac{dm}{dt}]$$

$$= m\frac{du}{dt} + u\left(\dot{m}_{\text{in}} - \dot{m}_{\text{out}}\right)$$

The following expressions describe the specific internal energy of the water, the rate of heat transfer out of the system to the surroundings and the mass of water in the system:

$$u = c(T - T_{\text{ref}})$$

$$Q_{\text{out}} = k(T - T_{\text{surr}})$$

$$m = m_0 + (\dot{m}_{\text{in}} - \dot{m}_{\text{out}})t$$

The differential equation describing the rate of change of the temperature reduces to:

$$\frac{dT}{dt} = \frac{\dot{m}_{\text{in}} c(T_{\text{in}} - T) - k(T - T_{surr})}{m_0 + (\dot{m}_{\text{in}} - \dot{m}_{\text{out}})t}$$

Where

T = Temperature

t = Time

m = Mass

U = Internal energy

u = Specific internal energy

Q = Heat transfer

c = Specific heat capacity

k = Heat transfer coefficient

$surr$ = Surroundings

ref = Reference zero state for u

Chapter 11

Law of Thermodynamics

First, what is temperature? Let's try to clear this first. Two systems in thermal contact should exchange no heat, on average, if and only if they are at the same temperature. Let's call the two systems S1 and S2. The combined system, treating S1 and S2 together, can be S3. The important question, consideration of which will lead us to a useful quantitative definition of temperature, is "How will the energy of S3 be distributed between S1 and S2?"

With a total energy E, S has many possible internal states (microstates). The atoms of S3 can share the total energy in many ways. Let's say there are N different states. Each state corresponds to a particular division of the total energy in the two subsystems S1 and S2. Many microstates can correspond to the same division, E1 in S1 and E2 in S2. A simple counting argument tells you that only one particular division of the energy, will occur with any significant probability. It's the one with the overwhelmingly largest number of microstates for the total system S3. That number, N(E1,E2) is just the product of the number of states allowed in each subsystem, N(E1,E2) = N1(E1)×N2(E2), and, since E1 + E2 = E, N(E1,E2) reaches a maximum when N1 × N2 is stationary with respect to variations of E1 and E2 subject to the total energy constraint.

For convenience, physicists prefer to frame the question in terms of the logarithm of the number of microstates N, and call this the entropy, S. You can easily see from the above analysis that two systems are in equilibrium with one another when $(dS/dE)_1 = (dS/dE)_2$, i.e., the rate of change of entropy, S, per unit change in energy, E, must be the same for both systems. Otherwise, energy will tend to flow from one subsystem to another as S3 bounces randomly from one microstate to another, the total energy E3 being constant, as the combined system moves towards a state of maximal total entropy. We define the temperature, T, by $1/T = dS/dE$, so that the equilibrium condition becomes the very simple $T_1 = T_2$.

This statistical mechanical definition of temperature does in fact correspond to your intuitive notion of temperature for most systems. So long as dS/dE is always positive, T is always positive. For common situations, like a collection of free particles, or particles in a harmonic oscillator potential, adding energy always increases the number of available microstates, increasingly faster with increasing total energy. So temperature increases with increasing energy, from zero, asymptotically approaching positive infinity as the energy increases.

ABSOLUTE ZERO

Absolute Zero is the lowest temperature theoretically possible according to kinetic theory, zero Kelvin (0 K), equivalent to –273.15°C/–459.67°F, at which molecules are in their lowest energy state. Although the third law of thermodynamics indicates the impossibility of reaching absolute zero, in practice temperatures of less than a billionth of a degree above absolute zero have been reached. Near absolute zero, the physical properties of some materials change substantially; for example, some metals lose their electrical resistance and become superconducting. Absolute zero is the point on the thermodynamic (absolute) temperature scale where the heat energy is at a minimum, that is, no more heat can be removed from the system.

According to classical physics this temperature would correspond to zero kinetic energy of the particles of the system, in the reference frame of the system's center of mass; this, however, is now known to be false—quantum mechanics explains why the energy of a system can never drop below its zero-point energy. By international agreement, the Celsius temperature scale starts at absolute zero with the value of exactly –273.15 °C (which approximately agrees with the original definition of the Celsius scale that was based on freezing and boiling point of water). It is approximately –459.67 °F on the Fahrenheit scale. In sciences, where SI units are in use, thermodynamic temperature is measured in kelvins where absolute zero is 0 K.

Scientists have made great advancements in achieving temperatures ever closer to absolute zero (where matter exhibits odd quantum effects). In 1994 the NIST achieved a record cold temperature of 700 nK (billionths of a kelvin). The absolute zero state was first proposed by Guillaume Amontons in 1702 who was investigating the relationship between pressure and temperature in gases. He lacked accurate and precise thermometers so his results were only semi-quantitative, but he established that the pressure of a gas increases by roughly one-third between "cold" temperatures and the boiling point of water. His work led him to speculate that a sufficient reduction in temperature would lead to the disappearance of pressure.

The problem is that all real gases liquify during the approach to absolute zero. According to kinetic theory, there should be no movement of individual molecules at absolute zero, so any material at this temperature would be solid. In a monatomic gas, most of the energy is in the form of translational motion, and the temperature can be measured in terms of the distribution of this motion, with slower speeds corresponding to lower temperatures, perhaps even down to absolute zero. But this is contrary to experimental evidence, as helium will never solidify at normal pressures, regardless of temperature. Because of quantum-mechanical effects, the speed at absolute zero is larger than zero and depends, along with the energy, on the volume within which a particle is confined.

At absolute zero, the molecules and atoms in a system are all in their ground state, the state of lowest possible energy, and a system has the least amount of kinetic energy allowed by the laws of physics. But the lowest possible zero-point energy for a confined particle in a box is not zero. Rather than being fixed and non-moving, the equation for the energy levels shows that no matter how low the temperature gets, even when the quantum number takes its minimum value of one, a particle still has some translational kinetic energy and motion. This is a reflection of Heisenberg's uncertainty principle, which states that the position and the momentum of a particle cannot both be known precisely at any given time. Similarly, using the harmonic approximation for the vibrations of a diatomic molecule, the quantum harmonic oscillator yields a positive zero-point energy even when the vibrational quantum number takes its minimum value of zero.

For polyatomic molecules, and for bodies such as crystals, whose normal mode motions can not be assigned to individual atoms or chemical bonds, the lowest-energy state is that of the system as a whole. Classically, the absolute temperature T of a system of molecules at thermodynamic equilibrium assigns an average of ½ kT to each quadratic kinetic and/or potential energy term in each mechanical degree of freedom, where k is Boltzmann's constant. But quantum mechanics shows that this is obeyed only for temperatures such that $kT > h\nu$, where h is Planck's constant and ν is a characteristic frequency. As T decreases, the assumption that energy is continuously variable fails whenever $h\nu$ exceeds kT. For vibrational modes in crystals, this happens at room temperature, which explains the deviation of the calculated specific heats of atomic crystals from the experimental Dulong-Petit law value of 3R /mole, a fact which puzzled late 19th century physicists and physical chemists.

At 0 K, (nearly) all molecular motion ceases and $\Delta S=0$ for any adiabatic process. Pure substances can (ideally) form perfect crystals as $T \rightarrow 0$. Planck's strong form of the third law of thermodynamics states that the entropy of a perfect crystal vanishes at absolute zero. However, if the lowest energy state is degenerate (more than one microstate), this cannot be true.

The original Nernst heat theorem makes the weaker and less controversial claim that the entropy change for any isothermal process approaches zero as $T \rightarrow 0$

$$\lim_{T \to 0} \Delta S = 0$$

which implies that the entropy of a perfect crystal simply approaches a constant value. The Nernst postulate identifies the isotherm T = 0 as coincident with the adiabat S = 0, although other isotherms and adiabats are distinct. As no two adiabats intersect, no other adiabat can intersect the T = 0 isotherm. Consequently no adiabatic process initiated at nonzero temperature can lead to zero temperature. An even stronger assertion is that It is impossible by any procedure to reduce the temperature of a system to zero in a finite number of operations. A perfect crystal is one in which the internal lattice structure extends uninterrupted in all directions.

The perfect order can be represented by translational symmetry along three (not usually orthogonal) axes. Every lattice element of the structure is in its proper place, whether it is a single atom or a molecular grouping. For substances which have two (or more) stable crystalline forms, such as diamond and graphite for carbon, there is a kind of "chemical degeneracy". The question remains whether both can have zero entropy at T = 0 even though each is perfectly ordered. Perfect crystals never occur in practice; imperfections, and even entire amorphous materials, simply get "frozen in" at low temperatures, so transitions to more stable states do not occur. Using the Debye model, the specific heat and entropy of a pure crystal are proportional to T^3, while the enthalpy and chemical potential are proportional to T^4. These quantities drop toward their T = 0 limiting values and approach with zero slopes. For the specific heats at least, the limiting value itself is definitely zero, as borne out by experiments to below 10 K. Even the less detailed Einstein model shows this curious drop in specific heats. In fact, all specific heats vanish as absolute zero, not just those of crystals.

Likewise for the coefficient of thermal expansion. Maxwell's relations show that various other quantities also vanish. These phenomena were unanticipated. Since the relation between changes in the Gibbs free energy, the enthalpy and the entropy is

$$\Delta G = \Delta H - T\Delta S$$

it follows that as T decreases, ΔG and ΔH approach each other (so long as ΔS is bounded). Experimentally, it is found that most chemical reactions are exothermic and release heat in the direction they are found to be going, toward equilbirum. That is, even at room temperature T is low enough so that the fact that $(\Delta G)_{T,P} < 0$ (usually) implies that $\Delta H < 0$. (In the opposite direction, each such reaction would of course absorb heat.) More than that, the slopes of the temperature derivatives of ΔG and ΔH converge and are equal to zero at T = 0, which ensures that ΔG and ΔH are nearly the same over a considerable range of temperatures, justifying the approximate empirical Principle of Thomsen and Berthelot, which says that the equilibrium state to which a system proceeds is the one which evolves the greatest amount of heat, i.e., an actual process is the most exothermic one.

As mentioned, absolute or thermodynamic temperature is conventionally measured in Kelvins (Celsius-size degrees), and increasingly rarely in the Rankine scale (Fahrenheit-size degrees). Absolute temperature is uniquely determined up to a multiplicative constant which specifies the size of the "degree", so the *ratios* of two absolute temperatures, T_2/T_1, are the same in all scales. The most transparent definition comes from the classical Maxwell-Boltzmann distribution over energies, or from the quantum analogs: Fermi-Dirac statistics (particles of half-integer spin) and Bose-Einstein statistics (particles of integer spin), all of which give the relative numbers of particles as (decreasing) exponential functions of energy over kT. On a macroscopic level, a definition can be given in terms of the efficiencies of "reversible" heat engines operating between hotter and colder thermal reservoirs.

THE THIRD LAW OF THERMODYNAMICS

A postulate related to but independent of the second law is that it is impossible to cool a body to absolute zero by any finite process. Although one can approach absolute zero as closely as one desires, one cannot actually reach this limit. The third law of thermodynamics, formulated by Walter Nernst and also known as the Nernst heat theorem, states that if one could reach absolute zero, all bodies would have the same entropy. In other words, a body at absolute zero could exist in only one possible state, which would possess a definite energy, called the zero-point energy. This state is defined as having zero entropy.

The third law of thermodynamics is usually stated as a definition:

"the entropy of a perfect crystal of an element at the absolute zero of temperature is zero".

At the absolute zero of temperature, there is zero thermal energy or heat. Since heat is a measure of average molecular motion, zero thermal energy means that the average atom does not move at all. Since no atom can have less than zero motion, the motion of every individual atom must be zero when the average molecular motion is zero. When none of the atoms which make up a perfectly ordered crystal move at all, there can be no disorder or different states possible for the crystal.

Taking the entropy of a perfect crystal of an element to be zero at the absolute zero of temperature establishes a method by which entropies of elements at any higher temperature can be determined. Since by definition $dS = q_{rev}/T$, the mathematical integral of dS from zero to any higher temperature T is the integral, over that temperature range, of q_{rev}/T. In other words. the difference $S - S^0$ is the integral from zero to a temperature T of $(C_p/T)dT$. The molar heat capacity at any temperature is a measurable quantity, and so this difference can be determined experimentally.

The third law of thermodynamics is simply the statement that S^0 is zero by definition for a pure element, and so if the heat capacity is measured under conditions of reversible heat flow, as it can be, and as a function of temperature at low temperatures, as it can be, then the entropy $S$of a pure element at any temperature T is given by:

S = the integral from zero to T of $(C_p/T)dT$

The values we usually write as S^0, the standard entropy of a substance, are actually the integral from zero to the standard temperature, 298.15 K, of $(C_p/T)dT$ in order that the values of standard entropies of substances can be used with standard values for other thermodynamic functions such as enthalpy and free energy.

The value of the entropy of an element at any temperature, including 298.15 K which is the temperature of the standard entropy S^0, can be obtained from careful measurements of the heat capacity of the element from the desired temperature down to absolute zero. Experimentally, chemists have been unable to reach the absolute zero of temperature, but measurements can be and have been made down to within 0.1 K of it and the heat capacity below measurable range can be accurately estimated. The entropy of the element is obtained from integration of the heat capacity measurements. If the crystal changes form, or melts, within the temperature range desired, then the entropy of that change must also be measured and included.

The entropies of compounds can be determined from thermodynamic measurements made on the reactions which form them from the elements or dissociate them to their constituent elements once the entropies of the elements themselves are established. The values of standard entropies of elements and compounds at 25°C given in the tables of the thermodynamic properties of pure substances are all based upon the third law of thermodynamics.

On a molar basis,

$$S = R \ln W$$

R = gas constant (units of J/mole. K)

W = the number of arrangements of the energy of atoms or molecules can take without changing the total energy in a system. This is also known as the *degeneracy*.

At absolute zero for a perfectly ordered crystalline substance, W = 1 (there is only one way to arrange the atoms in a perfect substance, hence there is only one way to distribute the energy)

This means that at T = 0, S = Rln(1) = 0.

The Third Law, then, gives us a method to find the absolute entropy for any substance: cool the substance to as close to absolute zero as possible and then measure the heat change as the substance is warmed up (remember, $\Delta S = q/T = S_{final} - S_{initial}$, but $S_{initial} = 0$ if we start at T = 0).

Absolute entropies for substances are tabulated at standard conditions.

The tabulated values of the absolute entropy allow us to calculate the change in entropy for chemical reactions:

$$\Delta S^o = \sum_{i=products} m_i S_i^o - \sum_{j=reactants} m_j S_j^o$$

m_i and m_j are the stoichiometric coefficients from the balanced reaction and S_i^o and S_j^o are the absolute entropies of the products and reactants.

ΔS^o will be positive if the chemical process leads to an increase in the number of ways to disperse energy.

This will happen if:

1. There are more moles of products than moles of reactants (there are more ways to distribute energy because there are more particles after reaction).
2. The number of moles of gaseous products is greater than the number of moles of gaseous reactants (gases can distribute energy in more ways because of the large mobility of the particles, so have a disproportionate effect on the change in entropy).
3. A change in phase from *s* to *l*, *l* to *g*, or *s* to *g*.

ABSOLUTE ENTROPIES

Energy values, as you know, are all relative, and must be defined on a scale that is completely arbitrary; there is no such thing as the absolute energy of a substance, so we can arbitrarily define the enthalpy or internal energy of an element in its most stable form at 298K and 1 atm pressure as zero.

The same is not true of the entropy; since entropy is a measure of the "dilution" of thermal energy, it follows that the less thermal energy available to spread through a system (that is, the lower the temperature), the smaller will be its entropy. In other words, as the absolute temperature of a substance approaches zero, so does its entropy.

This principle is the basis of the Third law of thermodynamics, which states that the entropy of a perfectly-ordered solid at 0° K is zero.

The absolute entropy of a substance at any temperature above 0° K must be determined by calculating the increments of heat q required to bring the substance from 0° K to the temperature of interest, and then summing the ratios q/T. Two kinds of experimental measurements are needed:

1. The enthalpies associated with any phase changes the substance may undergo within the temperature range of interest. Melting of a solid and vapourization of a liquid correspond to sizeable increases in the number of microstates available to accept thermal energy, so as these processes occur, energy will flow into a system, filling these new microstates to the extent required to maintain a constant temperature (the freezing or boiling point); these inflows of thermal energy correspond to the heats of fusion and vapourization. The entropy increase associated with melting, for example, is just $\Delta H_{fusion}/T_m$.
2. The heat capacity C of a phase expresses the quantity of heat required to change the temperature by a small amount ΔT, or more precisely, by an infinitessimal amount dT. Thus the entropy increase brought about by warming a substance over a range of temperatures that does not encompass a phase transition is given by the sum of the quantities C dT/T for each increment of temperature dT. This is of course just the integral

$$S_{0^0 \to T^0} = \int_{0^0}^{T^0} \frac{C}{T} dT$$

Because the heat capacity is itself slightly temperature dependent, the most precise determinations of absolute entropies require that the functional dependence of C on T be used in the above integral in place of a constant C. For rough determinations, one can simply plot the enthalpy increment C ΔT as a function of temperature and measure the area under the curve.

It is possible, though exceedingly difficult, to calculate the entropy of any system under any conditions of interest from the equation

$$S = R \ln W$$

It is also possible, using more advanced theoretical thermodynamics, to determine S experimentally by measuring heat capacities and enthalpies of phase transitions. Values of S determined experimentally, often referred to as "absolute" entropies, have been tabulated for many materials at many temperatures, and a few examples are given in table. We treat these values as observations and attempt to understand these in the context of equation.

Table: Absolute Entropies of Specific Substances

	T (°C)	S (J/mol°C)
$H_2O(g)$	25	188.8
$H_2O(l)$	25	69.9
$H_2O(l)$	0	63.3
$H_2O(s)$	0	41.3
$NH_3(g)$	25	192.4
$HN_3(l)$	25	140.6
$HN_3(g)$	25	239.0
$O_2(g)$	25	205.1
$O_2(g)$	50	207.4
$O_2(g)$	100	211.7

$CO(g)$	25	197.7
$CO(g)$	50	200.0
$CO_2(g)$	24	213.7
$CO_2(g)$	50	216.9
$Br_2(l)$	25	152.2
$Br_2(g)$	25	245.5
$I_2(s)$	25	116.1
$I_2(g)$	25	260.7
$CaF_2(s)$	25	68.9
$CaCl_2(s)$	25	104.6
$CaBr_2(s)$	25	130
$C_8H_{18}(s)$	25	361.1

There are several interesting generalities observed in table. First, in comparing the entropy of the gaseous form of a substance to either its liquid or solid form at the same temperature, we find that the gas always has substantially greater entropy. This is easy to understand from equation: the molecules in the gas phase occupy a very much larger volume. There are very many more possible locations for each gas molecule and thus very many more arrangements of the molecules in the gas. It is intuitively clear that W should be larger for a gas, and therefore the entropy of a gas is greater than that of the corresponding liquid or solid.

Second, we observe that the entropy of a liquid is always greater than that of the corresponding solid. This is understandable from our kinetic molecular view of liquids and solids. Although the molecules in the liquid occupy a comparable volume to that of the molecules in the solid, each molecule in the liquid is free to move throughout this entire volume. The molecules in the solid are relatively fixed in location. Therefore, the number of arrangements of molecules in the liquid is significantly greater than that in the solid, so the liquid has greater entropy by equation 1.

Third, the entropy of a substance increases with increasing temperature. The temperature is, of course, a measure of the average kinetic energy of the molecules. In a solid or liquid, then, increasing the temperature increases the total kinetic energy available to the molecules. The greater the energy, the more ways there are to distribute this energy amongst the molecules. Although we have previously only referred to the range of positions for a molecule as affecting W, the range of energies available for each molecule similarly affects W. As a result, as we increase the total energy of a substance, we increase W and thus the entropy.

Fourth, the entropy of a substance whose molecules contain many atoms is greater than that of a substance composed of smaller molecules. The more atoms there are in a molecule, the more ways there are to arrange those atoms. With greater internal flexibility, W is larger when there are more atoms, so the entropy is greater.

Fifth, the entropy of a substance with a high molecular weight is greater than that of substance with a low molecular weight. This result is a harder to understand, as it arises from the distribution of the momentum of the molecules rather than the positions and energies of the molecules. It is intuitively clear that the number of arrangements of the molecules is not affected by the mass of the molecules. However, even at the same temperature, the range of momentum available for a heavier molecule is greater than for a lighter one. To see why, recall that the momentum of a molecule is p=mv and the kinetic energy is $KE = mv^2/2 = p^2/2m$. Therefore, the maximum momentum available at a fixed total kinetic energy KE is:

$$p = \sqrt{2KE}$$

Since this is larger for larger mass molecules, the range of momentum is greater for heavier particles, thus increasing W and the entropy.

Helmholtz Free Energy

In thermodynamics, the Helmholtz free energy is a thermodynamic potential which measures the "useful" work obtainable from constant temperature, constant volume thermodynamic systems. For a simple system, with a fixed number of particles, the negative of the difference in the Helmholtz energy is equal to the maximum amount of work extractable from a thermodynamic process in which temperature is held constant. The Helmholtz free energy was developed by Hermann von Helmholtz and is denoted by the letter A or the letter F . The Helmholtz free energy as goes by the name Helmholtz energy or Helmholtz function among others.

The Helmholtz energy is defined as:

$$A = U - TS$$

where

- A is the Helmholtz energy (SI: joules, CGS: ergs),
- U is the internal energy of the system (SI: joules, CGS: ergs),
- T is the absolute temperature (kelvins),
- S is the entropy (SI: joules per kelvin, CGS: ergs per kelvin).

In earlier times the theory of heat, i.e. that heat is a form of energy having relation to vibratory motion, was beginning to supplant both the caloric theory, i.e. that heat is a fluid, and the four element theory in which heat was the lightest of the four elements. Many textbooks and teaching articles during these centuries presented these theories side by side.

Similarly, during these years, heat was beginning to be distinguished into different classification categorize, such as "free heat", "combined heat", "radiant heat", specific heat, heat capacity, "absolute heat", "latent caloric", "free" or "perceptible" caloric, among others. In 1780, for example, Laplace and Lavoisier stated: "In general, one can change the first hypothesis into the second by changing the words 'free heat, combined heat, and heat released' into 'vis viva, loss of vis viva, and increase of vis viva.'"

In this manner, the total mass of caloric in a body, called *absolute heat*, was regarded as a mixture of two components;

the free or perceptible caloric could affect a thermometer while the other component, the latent caloric, could not. The use of the words "latent heat" implied a similarity to latent heat in the more usual sense; it was regarded as chemically bound to the molecules of the body.

In the adiabatic compression of a gas, the absolute heat remained constant by the observed rise of temperature indicated that some latent caloric had become "free" or perceptible.During the early 19th century, the concept of perceptible or free caloric began to be referred to as "free heat" or heat set free. In 1824, for example, the French physicist Sadi Carnot, in his famous "Reflections on the Motive Power of Fire", speaks of quantities of heat 'absorbed or set free' in different transformations.

In 1882, the German physicist and physiologist Hermann von Helmholtz coined the phrase 'free energy' for the expression $E - TS$, in which the change in F (or G) determines the amount of energy 'free' for work under the given conditions. In modern use, we attach the term "free" to Gibbs free energy, i.e. for systems at constant pressure and temperature, or to Helmholtz free energy, i.e. for systems at constant volume and temperature, to mean 'available in the form of useful work.

With reference to the Gibbs free energy, we add the qualification that it is the energy free for non-volume work. To note, some books do not include the attachment "free", referring to G as simply Gibbs energy. This is the influence of a 1988 IUPAC meeting designed to unified terminologies between the USA, Europe, and other countries, in which descriptive 'free' was supposedly banished. This ruling, however, is still far from accepted and the majority of published articles and books still use the descriptive 'free' for both historical, informative, and for clarification reasons.

The Helmholtz free energy is defined by

$$A = U - TS \quad (1)$$

where E is the energy, T is the temperature, and S is the entropy. When a system changes its thermodynamic state, the change in Helmholtz free energy is therefore given by

$$dA = dU - TdS - SdT \quad (2)$$

If T and V are constant, the (2) reduces to

$$dA = dU - TdS$$

But the combined law of thermodynamics states that

$$dU - TdS + PdV \leq 0$$

and, since we have stipulated dV = 0, this becomes

$$dU - TdS \leq 0$$

Therefore

$$dA \leq 0$$

The Helmholtz free energy is intimately related to the equilibrium constant at constant volume A_V

$$A = -Kt \ln A_V$$

where *k* is Boltzmann's constant.

A system with fixed external parameters in thermal contact with a heat reservoir at equilibrium has a minimum Helmholtz free energy, commonly denoted A_0

GIBBS FREE ENERGY

Useful energy, or energy available to do work, is of central interest because there is never a shortage of useful work to be done, and usually an insufficiency of useful energy to do it. Chemical reactions can be made to do work, or to serve as pathways for the conversion of heat into work. Energy which is, or which can be, available to do useful work is called Gibbs free energy or more often simply free energy. Free energy is called Gibbs free energy and given the symbol G in honor of Josiah Willard Gibbs, a professor of chemistry at Harvard University who almost singlehandedly developed both the concept and the quantitative equations which describe it.

Free energy is a state function because it is formally defined only in terms of state functions, the state functions enthalpy and entropy, and the state variable temperature. The definition of free energy is:

$$G = H - TS$$

where (in SI units)

- G is the Gibbs energy (joules)
- H is the enthalpy (joules)
- T is the temperature (kelvins)
- S is the entropy (joules per kelvin)

A change in free energy dG would then be given by:

$$dG = dH - d(TS)$$

Since temperature is a state variable and entropy is a state function they are separable, and so:

$$dG = dH - TdS - SdT$$

However, at constant temperature the change in temperature dT is zero, so the term SdT is zero also. For any constant-temperature (isothermal) change, then $dG = dH - TdS$. Rearrangement of this equation gives, for larger changes, the Gibbs-Helmholtz equation:

$$DH = DG + TDS$$

Total energy available as heat = DG + energy not available for doing work

It therefore follows that DG must be the energy which is available for doing work. This free or available energy which can do useful work is then a most useful thing to know or to have.

The conventions of heat flow q are also those of enthalpy change. When DH is negative, heat is evolved (or given off) in the exothermic process or reaction, while when DH is positive, heat is absorbed in the endothermic process or reaction from the surroundings. Free energy change also follows the same conventions. When DG is negative, energy is evolved (given up) in the process or reaction and the change can do useful work on something else. This is a spontaneous process, and is sometimes called an exergonic (energy-giving) process. When DG is positive, energy is absorbed in the process or reaction and useful work must be done on the materials so that the reaction will occur. This is a nonspontaneous process, and is sometimes called an endergonic (energy-requiring) process.

Derivation

Let S_{tot} be the total entropy of an isolated system, that is, a system which cannot exchange heat or mass with its surroundings. By the second law of thermodynamics:

$$S\Delta S_{tot} \geq 0$$

and if $\Delta S_{tot} = 0$ then the process is reversible.

The heat transfer Q vanishes for an adiabatic system. Any adiabatic process is reversible and isentropic.

$$\left(\frac{\Delta q}{T} = \Delta S = 0\right)$$

Now consider diabatic systems, having internal entropy S_{int}. Such a system is thermally connected to its surroundings, which have entropy S_{ext}.

The entropy form of the second law does not apply directly to the diabatic system, it only applies to the closed system formed by both the system and its surroundings. Therefore a process is possible if

$$\Delta S_{int} + \Delta S_{ext} \geq 0 .$$

We will try to express the left side of this equation entirely in terms of state functions. ΔS_{ext} is defined as:

$$DS_{ext} = -\frac{\Delta q}{T}$$

Temperature T is the same for two systems in thermal equilibrium. (By the zeroeth law of thermodynamics, if a system is in thermal equilibrium with a second and a third system, the latter two are in equilibrium as well.) Also, Δq_{rev} is heat transferred *to* the system, so $-\Delta q_{rev}$ is heat transferred to the surroundings, and $\Delta Q/T$ is entropy gained by the surroundings. We now have:

$$\Delta S_{\text{int}} - \frac{\Delta q}{T} \geq 0$$

Multiply both sides by T:

$$T\Delta S_{\text{int}} - \Delta q \geq 0$$

ΔQ is heat transferred *to* the system; if the process is now assumed to be isobaric, then $\Delta q_p = \Delta H$:

$$T\Delta S_{\text{int}} - \Delta H \geq 0$$

ΔH is the enthalpy change of reaction (for a chemical reaction at constant pressure and temperature). Then

$$\Delta H - T\Delta S_{\text{int}} \geq 0$$

for a possible process. Let the change ΔG in Gibbs energy be defined as

$$\Delta G = \Delta H - T\Delta S_{\text{int}} \quad (1)$$

Notice that it is not defined in terms of any external state functions, such as ΔS_{ext} or ΔS_{tot}. Then the second law becomes:

$\Delta G < 0$ favored reaction

$\Delta G = 0$ reversible reaction

$\Delta G > 0$ disfavored reaction

Also, the sign of Delta G tells us about the spontaneity of the reaction.

$\Delta G < 0$ Spontaneous

$\Delta G = 0$ Equilibrium

$\Delta G > 0$ Nonspontaneous

Gibbs energy G itself is defined as

$$G = H - T\,S_{\text{int}} \quad (2)$$

but notice that to obtain equation (2) from equation (1) we must assume that T is constant.

Thus, Gibbs energy is most useful for thermochemical processes at constant temperature and pressure: both isothermal and isobaric. Such processes don't move on a *P-T* diagram, such as phase change of a pure substance, which takes place at the saturation pressure and temperature. However, chemical reactions do undergo changes in chemical potential, which is a state function. Thus, thermodynamic processes are not confined to the two dimensional *P-V* diagram. There is a third dimension for *n*, the quantity of gas. Naturally for the study of explosive chemicals, the processes are not necessarily isothermal and isobaric. For these studies, Helmholtz free energy is used.

Use of Gibbs Free Energy: Phase Transitions

We are interested in the cases where two phases of the same substance can coexist. An example is ice in water at 0ºC: if no heat is exchanged with the surroundings, the mixture will persist indefinitely.

In practical cases we are interested in charting the behaviour of the substance as a function of external temperature and pressure. For most choices, only one phase will persist at equilibrium - water, or ice, or vapour, for example. But for certain choices, two or even three phases may coexist.

The phase diagram for water is shown below.

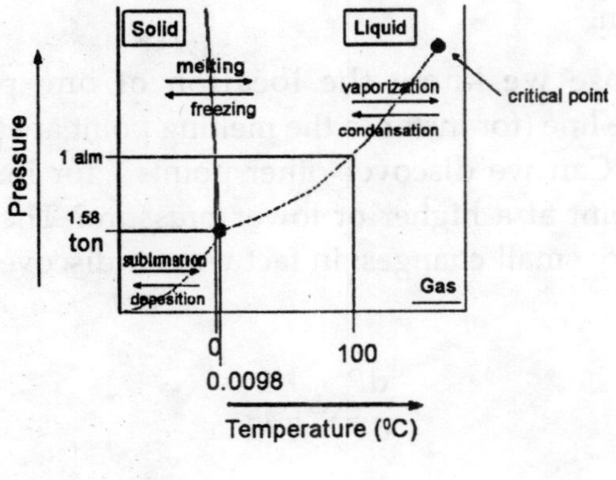

It differs from a generic substance in the backward slope of the ice-water coexistence line: putting ice under pressure may cause it to melt, whereas most liquids close to their freezing point will freeze under pressure.

What determines the lines of phase coexistence?

Consider a system of two phases, 1 (say gas) and 2 (say liquid) at a fixed pressure and temperature.

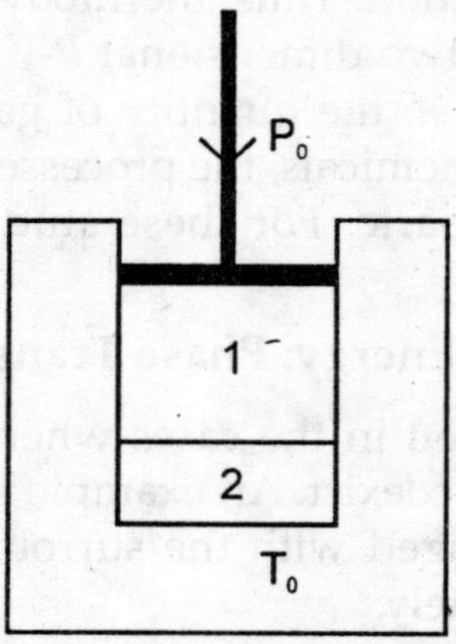

From the fact that the system will minimize its Gibbs free energy, we can derive the following condition for phase coexistence: The Gibbs free energy per unit mass in each phase is the same.

At temperatures and pressures not on the coexistence line the Gibbs free energy of one phase will be lower than the other, and only the phase with the lower energy will be present at equilibrium.

Suppose we know the location of one point on a coexistence line (for instance the melting point at atmospheric pressure). Can we discover other points - for instance, the melting point at a higher or lower pressure? The answer is yes, for very small changes: in fact we can discover the slope of the line:

$$\frac{dP}{dT} = \frac{L}{T\Delta V}$$

where *L* is the latent heat of the transition for some quantity of the substance, and ΔV is the volume change of the same quantity as it undergoes the transition. This is called the Clausius-Clapeyron equation.

We can use this equation to estimate the boiling point of water at the top of Mount Everest, where the pressure is only about a third of that at sea-level, to be 82°C.

(Input data: pressure at the top of Everest is 0.36 atmospheres, the density of water vapour at 100ºC is 0.598 kg/m^3 and the latent heat is 2.257 × 103 J/g.)

Of course this estimate requires assuming that the phase coexistence line has negligible curvature over a fairly large range, so it is no surprise that the result is only approximate. By looking at the phase diagram above you should be able to say if the true temperature is higher or lower.

HEAT CAPACITY

The heat capacity of a substance is a measure of how well the substance stores heat. Whenever we supply heat to a material, it will necessarily cause an increase in the material's temperature. The heat capacity is defined as the amount of heat required per unit increase in temperature, so that

Heat added = heat capacity × (change in temperature)

Thus, materials with large heat capacities, like water, hold heat well - their temperature won't rise much for a given amount of heat - whereas materials with small heat capacities, like copper, don't hold heat well - their temperature will rise significantly when heat is added.

Heat capacity (usually denoted by a capital C, often with subscripts) is a measurable physical quantity that characterizes the ability of a body to store heat as it changes in temperature. It is defined as the rate of change of temperature as heat is added to a body at the given conditions and state of the body (foremost its temperature).

In the International System of Units, heat capacity is expressed in units of joules per kelvin. It is termed an "extensive quantity" because it is sensitive to the size of the object (for example, a bathtub of water has a greater heat capacity than a cup of water). Dividing heat capacity by the body's mass yields a specific heat capacity (also called more properly "mass-specific heat capacity" or more loosely "specific heat"), which is an "intensive quantity," meaning it is no longer dependent on amount of material, and is now more dependent on the type of material, as well as the physical conditions of heating.

Heat capacity is mathematically defined as the ratio of a small amount of heat δQ added to the body, to the corresponding small increase in its temperature dT:

$$C = \left(\frac{\delta Q}{dT}\right)_{cond.} = T\left(\frac{dS}{dT}\right)_{cond.}$$

For thermodynamic systems with more than one physical dimension, the above definition does not give a single, unique quantity unless a particular infinitesimal path through the system's phase space has been defined (this means that one needs to know at all times where all parts of the system are, how much mass they have, and how fast they are moving). This information is used to account for different ways that heat can be stored as kinetic energy (energy of motion) and potential energy (energy stored in force fields), as an object expands or contracts.

For all real systems, the path though these changes must be explicitly defined, since the value of heat capacity depends on which path from one temperature to another, is chosen. Of particular usefulness in this context are the values of heat capacity for constant volume, CV, and constant pressure, CP. These will be defined below.

Heat Capacity of Fluids

The state of a simple fluids with fixed mass is described by two thermodynamic parameters such as temperature T and pressure P. Therefore as mentioned above, one may distinguish between heat capacity at constant volume, C_V, and heat capacity at constant pressure, C_P:

$$C_V = \left(\frac{\delta Q}{dT}\right)_V = T\left(\frac{\partial S}{\partial T}\right)_V$$

$$C_P = \left(\frac{\delta Q}{dT}\right)_P = T\left(\frac{\partial S}{\partial T}\right)_P$$

where δQ is the infinitesimal amount of heat added, and dT is the subsequent rise in temperature.

The increment of internal energy is the heat added and the work added:

$$dU = T\,dS - P\,dV$$

So the heat capacity at constant volume is

$$C_V = \left(\frac{\partial U}{\partial T}\right)_V$$

The enthalpy is defined by H = U + PV. The increment of enthalpy is

$$dH = dU + (PdV + VdP)$$

which at dP=0, reduces to:

$$dH = T\,dS + V\,dP$$

So the heat capacity at constant pressure is

$$C_P = \left(\frac{\partial H}{\partial T}\right)_P$$

Note that this last "definition" is a bit circular, since the concept of "enthalpy" itself was invented to be a measure of heat absorbed or produced at constant pressures (the conditions in which chemists usually work). As such, enthalpy merely accounts for the extra heat which is produced or absorbed by pressure-volume work at constant pressure. Thus, it is not surprising that constant-pressure heat capacities may be defined in terms of enthalpy, since "enthalpy" was defined in the first place to make this so.

Specific Heat Capacity

The specific heat capacity of a material is

$$c = \frac{\partial C}{\partial m}$$

which in the absence of phase transitions is equivalent to

$$c = c_m = \frac{C}{m} = \frac{C}{\rho V}$$

- C is the heat capacity of a body made of the material in question ($J{\cdot}K^{-1}$)
- *m* is the mass of the body (kg)
- *V* is the volume of the body (m^3)
- $\delta = mV^{-1}$ is the density of the material ($kg{\cdot}m^{-3}$)

For gases, and also for other materials under high pressures, there is need to distinguish between different boundary conditions for the processes under consideration (since values differ significantly between different conditions). Typical processes for which a heat capacity may be defined include isobaric (constant pressure, dP = 0) and isochoric (constant volume, dV = 0) processes, and one conventionally writes for gases:

$$c_P = \left(\frac{\partial C}{\partial m}\right)_P$$

$$c_V = \left(\frac{\partial C}{\partial m}\right)_V$$

Units shown are SI units but, of course, any consistent set of units may be used.

A related parameter to c is CV–1, the volumetric heat capacity, (J·m–3·K–1 in SI units). In engineering practice, cV for solids or liquids often signifies a volumetric heat capacity, rather than a constant-volume one. In such cases, the mass-specific heat capacity (specific heat) is often explicitly written with the subscript m, as cm. Of course, from the above relationships, for solids one writes:

$$c_m = \frac{C}{m} = \frac{c_V}{\rho}$$

Dimensionless Heat Capacity

The dimensionless heat capacity of a material is

$$C^* = \frac{C}{nR} = \frac{C}{Nk}$$

where

- C is the heat capacity of a body made of the material in question ($J \cdot K^{-1}$)
- n is the amount of matter in the body (mol)
- R is the gas constant ($J \cdot K^{-1} \cdot mol^{-1}$)
- nR=Nk is the amount of matter in the body ($J \cdot K^{-1}$)
- N is the number of molecules in the body. (dimensionless)
- k is Boltzmann's constant ($J \cdot K^{-1} \cdot molecule^{-1}$) .

Gas Phase

According to the equipartition theorem from classical statistical mechanics, for a system made up of independent and quadratic degrees of freedom, any input of energy into a closed system composed of N molecules is evenly divided among the degrees of freedom available to each molecule. It can be shown that, in the classical limit of statistical mechanics, for each independent and quadratic degree of freedom, that

$$E_i = \frac{k_B T}{2}$$

where

E_i is the mean energy (measured in joules) associated with degree of freedom i.

T is the temperature (measured in Kelvin)

k_B is Boltzmann's constant, ($1.380\ 6505(24) \times 10^{-23}$ J K^{-1})

In the case of a monatomic gas such as helium under constant volume, if it assumed that no electronic or nuclear quantum excitations occur, each atom in the gas has only 3 degrees of freedom, all of a translational type. No energy dependence is associated with the degrees of freedom which define the position of the atoms. While, in fact, the degrees of freedom corresponding to the momenta of the atoms are quadratic, and thus contribute to the heat capacity. There are N atoms, each of which has 3 components of momentum, which leads to 3N total degrees of freedom. This gives:

$$C_V = \left(\frac{\partial U}{\partial T}\right)_V = \frac{3}{2} N k_B = \frac{3}{2} n R$$

$$C_V, m = \frac{C_V}{n} = \frac{3}{2} R = 1.5R$$

where

C_V is the heat capacity at constant volume of the gas

$C_{V,m}$ is the molar heat capacity at constant volume of the gas

N is the total number of atoms present in the container

n is the number of moles of atoms present in the container (n is the ratio of N and Avogadro's number)

R is the ideal gas constant, (8.314570[70] J $K^{-1}mol^{-1}$). R is equal to the product of Boltzmann's constant k_B and Avogadro's number

In the somewhat more complex case of an ideal gas of diatomic molecules, the presence of internal degrees of freedom are apparent. In addition to the three translational degrees of freedom, there are rotational and vibrational degrees of freedom. In general, the number of degrees of freedom, f, in a molecule with n_a atoms is $3n_a$:

$$f = 3n_a$$

Mathematically, there are a total of three rotational degrees of freedom, one corresponding to rotation about each of the axes of three dimensional space. However, in practice we shall only consider the existence of two degrees of rotational freedom for linear molecules. This approximation is valid because the moment of inertia about the internuclear axis is vanishingly small with respect other moments of inertia in the molecule (this is due to the extremely small radii of the atomic nuclei, compared to the distance between them in a molecule). Quantum mechanically, it can be shown that the interval between successive rotational energy eigenstates is inversely proportional to the moment of inertia about that axis. Because the moment of inertia about the internuclear axis is vanishingly small relative to the other two rotational axes, the energy spacing can be considered so high that no excitations of the rotational state can possibly occur unless the temperature is extremely high. We can easily calculate the expected number of vibrational degrees of freedom (or vibrational modes). There are three degrees of translational freedom, and two degrees of rotational freedom, therefore

$$f_{vib} = f - f_{trans} - f_{rot} = 6 - 3 - 2 = 1$$

Each rotational and translational degree of freedom will contribute R/2 in the total molar heat capacity of the gas. Each vibrational mode will contribute R to the total molar heat capacity, however. This is because for each vibrational mode, there is a potential and kinetic energy component. Both the potential and kinetic components will contribute R/2 to the total molar heat capacity of the gas. Therefore, we expect that a diatomic molecule would have a molar constant-volume heat capacity of

$$C_{V,m} = \frac{3R}{2} + R + R = \frac{7R}{2} = 3.5R$$

where the terms originate from the translational, rotational, and vibrational degrees of freedom, respectively. However, as the atoms composing the molecules become heavier, the heat capacities move closer to their expected values. One of the reasons for this phenomenon is the quantization of vibrational, and to a lesser extent, rotational states. In fact, if it is assumed that the molecules remain in their lowest energy vibrational state because the inter-level energy spacings are large, the predicted molar constant volume heat capacity for a diatomic molecule becomes

$$C_{V,m} = \frac{3R}{2} + R = \frac{5R}{2} = 2.5R$$

which is a fairly close approximation of the heat capacities of the lighter molecules in the above table. If the quantum harmonic oscillator approximation is made, it turns out that the quantum vibrational energy level spacings are actually inversely proportional to the square root of the reduced mass of the atoms composing the diatomic molecule. Therefore, in the case of the heavier diatomic molecules, the quantum vibrational energy level spacings become finer, which allows more excitations into higher vibrational levels at a fixed temperature.

Solid Phase

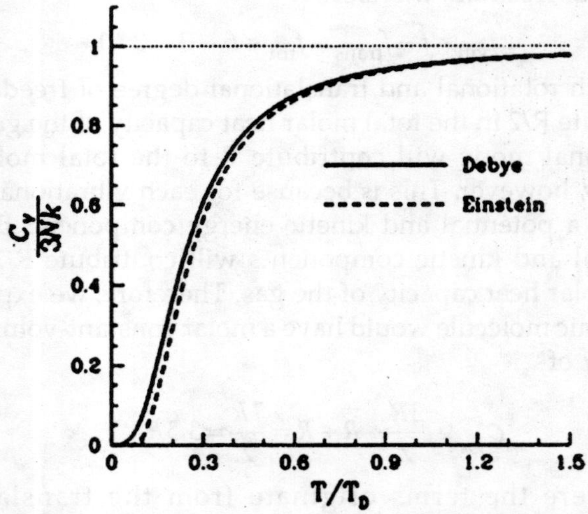

The dimensionless heat capacity divided by three, as a function of temperature as predicted by the Debye model and by Einstein's earlier model. The horizontal axis is the temperature divided by the Debye temperature. Note that, as expected, the dimensionless heat capacity is zero at absolute zero, and rises to a value of three as the temperature becomes much larger than the Debye temperature. The red line corresponds to the classical limit of the Dulong-Petit law

For matter in a crystalline solid phase, the Dulong-Petit law, which was discovered empirically, states that the dimensionless specific heat capacity assumes the value 3. Indeed, for solid metallic chemical elements at room temperature, heat capacities range from about 2.8 to 3.4 (beryllium being a notable exception at 2.0).

The theoretical maximum heat capacity for larger and larger multi-atomic gases at higher temperatures, also approaches the Dulong-Petit limit of 3R, so long as this is calculated per mole of atoms, not molecules. The reason is that gases with very large molecules, in theory have almost the same high-temperature heat capacity as solids, lacking only the (small) heat capacity contibution that comes from potential energy that cannot be stored between separate molecules in a gas.

The Dulong-Petit "limit" results from the equipartition theorem, and as such is only valid in the classical limit of a microstate continuum, which is a high temperature limit. For light and non-metallic elements, as well as most of the common molecular solids based on carbon compounds at standard ambiant temperature, quantum effects may also play an important role, as they do in multi-atomic gases. These effects usually combine to give heat capacities lower than 3 R per mole of atoms in the solid, although heat capacities calculated per mole of molecules in molecular solids may be more than 3 R. For example, the heat capacity of water ice at the melting point is about 4.6 R per mole of molecules, but only 1.5 R per mole of atoms. The lower number results from the "freezing out" of possible vibration modes for light atoms at suitably low temperatures, just as in many gases. These effects are seen in solids more often than liquids: for example the heat capacity of liquid water is again close to the theoretical 3 R/mole of atoms of the Dulong-Petit theoretical maximum.

Heat Capacity at Absolute Zero

From the definition of entropy

$$TdS = \delta Q$$

we can calculate the absolute entropy by integrating from zero temperature to the final temperature T_f

$$S(T_f) = \int_{T=0}^{T_f} \frac{\delta Q}{T} = \int_0^{T_f} \frac{\delta Q}{dT}\frac{\delta T}{T} = \int_0^{T_f} C(T)\frac{dT}{T}$$

The heat capacity must be zero at zero temperature in order for the above integral not to yield an infinite absolute entropy, thus violating the third law of thermodynamics. One of the strengths of the Debye model is that (unlike the preceding Einstein model) it predicts an approach of heat capacity toward zero as zero temperature is approached, and also predicts the proper mathematical form of this approach.

STANDARD STATES

Changes in energy content of a system are most easily compared if everyone agrees on a set of reference conditions. There are several conventions, but the commonly used set refers changes in chemical systems to defined standard states. The primary reference is to the standard states of the elements.

The standard state and the zero reference energy level

Elements in their standard states are considered to have chemical potentials and enthalpies of 0, or

$$\mu^{o}{}_{element} = 0$$

The standard state of an element is its natural state at 1 atms pressure, 25°C.

By defining the free energy of the elements in this way, we can regard any compound as having a chemical potential (partial free enrgy), or an enthalpy of formation, composed of the sum of all changes in chemical potential (or of enthalpy) for the reactions leading to its formation, by any convenient path. Since free energy and enthalpy are variables of state, the value is a unique function of the state, so this approach can be used to define the the relative energy content of any chemical system by reference to the work needed to get there starting from the elements.

Standard states of Solutes and Gases, and Free Energy Changes of Reaction

In order to compare free energies for chemical processes, it is convenient to normalize free energy changes so as to eliminate differences in reaction volume. This is achieved by using ΔG^o, the molar free energy change, and a standard state for reactants in solution.

Under normal temperature and pressure (NTP) of 25°C and1 atms pressure:

Solutes are in their standard states when they have an activity of 1 M.

Gases are in their standard states when they are at a pressure of 1 atms.

Standard state of the Solvent

Treatment of the solvent represents a tricky issue. By convention, the solvent is assumed to have a standard state of 1 M (the units being necessary to avoid dimensions in logaritmic terms), which does not change under the conditions of biochemical reaction (solutes at dilute solution). In the case of aqueous solutions, this may seem odd, because the activity of liquid water is 55.35 M (at NTP). This convention is adopted because the interactions with the solvent are subsumed under the standard chemical potentials of the solutes (which are referred to the standard state of a 1 M solution), and under the activity coefficients relating activities to concentrations. For reactions, the interactions are subsumed under ΔG^o, since this is derived by summation of values for μ^o. The convention allows one to ignore interactions between solute and solvent; this is usually appropriate since they are not significantly affected by changes in the solvent. In effect, for biochemical reactions in which water is a substrate or product (mainly

hydrolysis and lyase reactions), water is omitted from the thermodynamic equation. Thus for ATP hydrolysis:

$$\text{ATP} + \text{H}_2\text{O} \Leftrightarrow \text{ADP} + \text{Pi}$$

$$\Delta G' = \Delta G^{0'} + RT \ln \frac{(ADP).(Pi)}{(ATP)}$$

Chapter 12

Compressible Flow

A compressible flow is a situation in which the density of the flow cannot be assumed to be constant. In general, this is the case where the Mach number in part or all of the flow approaches or exceeds 0.3 (This is rather arbitrary but is a rule of thumb generally used). It can be proven that for gas flows with Mach number less than 0.3 the change in density is less than 5per cent.

While the above definition for compressible flow holds for most of the cases, it is possible for have flows with varying density and the flow is still incompressible. For example, consider a flow of water in oceans or air flow in atmosphere. The density varies as we move to different heights, but the flow can still be treated as incompressible. The factor that distinguishes a flow from being compressible or incompressible is the fact that in compressible flow the changes in the velocity of the flow can lead to changes the temperature which are not negligible. One the other hand in case of incompressible flow, the changes in the internal energy (i.e. temperature) are negligible even if the the entire kinetic energy of the flow is converted to internal energy (i.e. the flow is brought to rest).

These definitions, though they seem to be inconsistent, are all saying one and the same thing: the Mach number of the flow is high enough so that the effects of compressibility can no longer be neglected.

For subsonic compressible flows, it is sometimes possible to model the flow by applying a correction factor to the answers derived from incompressible calculations or modelling - for example, the Glauert-Prandtl rule

$$\frac{a_c}{a_i} \sim \frac{1}{\sqrt{1-M^2}}$$

(a_c is compressible lift curve slope, a_i is the incompressible lift curve slope, and M is the Mach number).

For many other flows, their nature is qualitatively different to subsonic flows. A flow where the local Mach number reaches or exceeds 1 will usually contain shock waves. A shock is an abrupt change in the velocity, pressure and temperature in a flow; the thickness of a shock scales with the molecular mean free path in the fluid (typically a few micrometers).

Shocks form because information about conditions downstream of a point of sonic or supersonic flow can not propagate back upstream past the sonic point.

The behaviour of a fluid changes radically as it starts to move above the speed of sound (in that fluid). For example, in subsonic flow, a stream tube in an accelerating flow contracts. But in a supersonic flow, a stream tube in an accelerating flow expands. To interpret this in another way, c onsider steady flow in a tube that has a sudden expansion: the tube's cross section suddenly widens, so the cross-sectional area increases.

In subsonic flow, the fluid speed drops after the expansion (as expected). In supersonic flow, the fluid speed increases. This sounds like a contradiction, but it isn't: the mass flux is conserved but because supersonic flow allows the density to change, the volume flux is not constant.

Compressible flow appears in many natural and many technological processes. Compressible flow deals with more than air, including steam, natural gas, nitrogen and helium, etc. For instance, the flow of natural gas in a pipe system, a common method of heating, should be considered a compressible flow. These processes include the flow of gas in the exhaust system of an internal combustion engine, and also gas turbine, a problem that led to the Fanno flow model. The above flows that were mentioned are called internal flows.

Compressible flow also includes flow around bodies such as the wings of an airplane, and is considered an external flow.

These processes include situations not expected to have a compressible flow, such as manufacturing process such as the die casting, injection molding. The die casting process is a process in which liquid metal, mostly aluminium, is injected into a mold to obtain a near final shape. The air is displaced by the liquid metal in a very rapid manner, in a matter of milliseconds, therefore the compressibility has to be taken into account.

Clearly, Engineers are not the only ones who have to deal with some aspect of compressible flow. For manufacturing engineers there are many situations where the compressibility or compressible flow understating is essential for adequate design. For instance, the control engineers who are using pneumatic systems use compressed substances. The cooling of some manufacturing systems and design of refrigeration systems also utilizes compressed air flow knowledge. Some aspects of these systems require consideration of the unique phenomena of compressible flow.

The compressible flow is a subset of fluid mechanics/ hydraulics and therefore the knowledge development followed the understanding of incompressible flow. Early contributors were motivated from a purely intellectual curiosity, while most later contributions were driven by necessity. As a result, for a long time the question of the speed of sound was bounced around.

BASIC EQUATIONS OF COMPRESSIBLE FLUID FLOW

Continuity Equation

Fluids, by definition can flow, but are essentially incompressible. This provides some very useful information about how fluids behave when they flow through a pipe, or a hose. Consider a hose whose diameter decreases along its length, as shown in the Figure below. The "continuity equation" is a direct consequence of the rather trivial fact that what goes into the hose must come out. The volume of water flowing through the hose per unit time (i.e. the flow rate at the left must be equal to the flow rate at the right or in fact anywhere along the hose. Moreover, the flow rate at and point in the hose is equal to the area of the hose at that point times the speed with which the fluid is moving:

flow rate = (area) × (velocity)

You can easily verify that (area)x(velocity) has units m^3/t which is correct for volume per unit time.

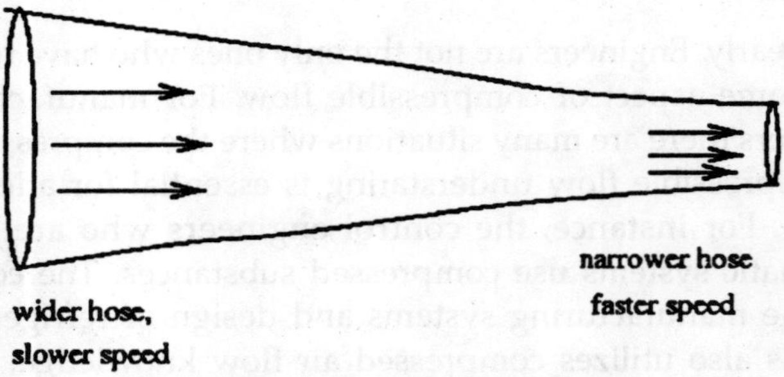

Fig. Fluid flow in a hose of variable size

These considerations lead us directly to the continuity equation, which states that

Area × velocity = constant

everywhere along the hose. This has the important consequence that as the area of the hose decreases, the velocity of the fluid must increase, in order to keep the flow rate constant. Anyone who has pinched one end of a garden hose has experienced this effect: the smaller you pinch the end of the hose, the faster the water comes out.

A continuity equation expresses a conservation law by equating a net flux over a surface with a loss or gain of material within the surface. Continuity equations often can be expressed in either integral or differential form.

Let's start with a small, fixed volume of fluid somewhere in the middle of a flow stream. This elemental volume has sides of lengths Δx, Δy and Δz.

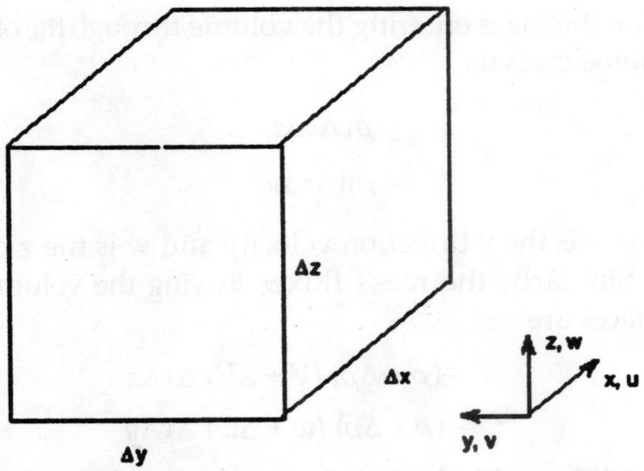

Fig. Illustration of the elemental volume used to derive the equations.

These lengths are short enough so that changes in all fluid properties across the volume may be well approximated with linear functions. On the other hand, these dimensions must me large enough so that the fluid may be considered as a continuum (i.e., much larger than the molecular scale). The mass of fluid in this elemental volume depends on the amount of fluid entering and leaving through the faces. The difference between these two is the rate of mass that accumulates in the volume. The rate of mass entering a face is the product of the density, the fluid velocity and the face area. For example, on the side facing the reader, the density (ρ) is multiplied by the velocity in the x direction (u) and the area of the face Δy Δz. Thus, the mass flux entering the volume through this face is

$$\text{mass flux in} = \rho u \Delta y \Delta z$$

The mass leaving the volume on the opposite side of the volume is again the product of density, velocity and area, but the density and velocity may have changed as the fluid passed through the volume. We will express these changes as small quantities (since our volume is small enough), i.e., $\rho + \Delta\rho$ and $u + \Delta u$. The mass flux leaving that face is thus

$$\text{mass flux out} = -(\rho + \Delta\rho)\ (u + \Delta u)\ \Delta y \Delta z$$

The negative sign tells us that the mass is leaving, rather than entering, the control volume. Performing the same analysis on the mass entering the volume through the other faces of the volume gives us

$$\rho v \Delta x \Delta z$$

$$\rho w \Delta x \Delta y$$

where v is the y direction velocity and w is the z direction velocity. Similarly, the mass fluxes leaving the volume on the opposite faces are

$$-(\rho + \Delta\rho)\ (V + \Delta V)\ \Delta x \Delta z$$

$$-(\rho + \Delta\rho)\ (w + \Delta w)\ \Delta x \Delta y$$

All of these added together must equal the mass of fluid accumulating in the volume, $\Delta x \Delta y \Delta z$,

$$\left(\frac{\Delta\rho}{\Delta t}\right)\Delta x \Delta y \Delta z$$

Putting all of these together, we have

$$\left(\frac{\Delta\rho}{\Delta t}\right)\Delta x \Delta y \Delta z = \rho u \Delta y \Delta z + \rho v \Delta x \Delta z + \rho w \Delta x \Delta y$$
$$-(\rho+\Delta\rho)(u+\Delta u)\Delta y \Delta z - (\rho+\Delta\rho)(v+\Delta v)\Delta x \Delta z$$
$$-(\rho+\Delta\rho)(w+\Delta w)\Delta x \Delta y$$

Multiplying out the quantities in parentheses results in the cancellation of some terms and the appearance of higher-order terms such as Δρ Δu Δy Δz. Since the quantities preceded by Δ are very small, products of these quantities will be extremely small, depending on the number of Δ terms included in the product. The terms with four of these will be much smaller than the terms with only three Δ terms. Thus, all higher order terms are neglected. This leaves

$$\left(\frac{\Delta\rho}{\Delta t}\right)\Delta x \Delta y \Delta z = -(u\Delta\rho + \rho\Delta u)\Delta y \Delta z - (\rho\Delta v + v\Delta\rho)\Delta x \Delta z - (\rho\Delta w + w\Delta\rho)\Delta x \Delta y$$

or

$$\left(\frac{\Delta\rho}{\Delta t}\right)\Delta x \Delta y \Delta z = -\Delta(\rho u)\Delta y \Delta z - \Delta(\rho v)\Delta x \Delta z - \Delta(\rho w)\Delta x \Delta y$$

which, when divided by ΔxΔyΔz and rearranged, yields

$$\frac{\partial \rho}{\partial t} + \frac{\partial(\rho u)}{\partial x} + \frac{\partial(\rho v)}{\partial y} + \frac{\partial(\rho w)}{\partial z} = 0$$

The application of basic calculus (taking the limit as Δt tends to 0) allows us to write this equation as

$$\frac{\partial \rho}{\partial t} + \frac{\partial(\rho u)}{\partial x} + \frac{\partial(\rho v)}{\partial y} + \frac{\partial(\rho w)}{\partial z} = 0$$

where the symbol $\partial / \partial t$, for example, is a "partial derivative" with respect to time. Partial derivatives are used when the function (velocity or density in this case) depends on several variables.

The Continuity Equation may be simplified for some common flow situations as follows. If the fluid may be treated as incompressible (as is the case with water or in low velocity air flows), the density will be constant. The Continuity Equation then becomes

$$\frac{\partial u}{\partial x} + \frac{\partial v}{\partial y} + \frac{\partial w}{\partial z} = 0 \quad \text{(incompressible fluid)}$$

In the case when the flow is steady (all time derivatives are zero), then

$$\frac{\partial(\rho u)}{\partial x} + \frac{\partial(\rho v)}{\partial y} + \frac{\partial(\rho w)}{\partial z} = 0$$

Note that in this equation, the density and velocities are still functions of the spatial coordinates x,y and z.

Momentum (Navier Stokes) Equations

Again we start with a small, fixed volume of fluid somewhere in the middle of a flow stream with sides of lengths Δx, Δy and Δz. The law of the conservation of momentum states that the rate of change of momentum in the control volume must equal the net momentum flux into the control volume plus any external forces acting on the control volume. We will first deal with the momentum change and flux, then with the external forces. Recall that momentum, the product of mass and velocity, is a vector quantity. This derivation will be based on the momentum in the x direction in Cartesian coordinates. Similar derivations may be demonstrated for the y and z direction. This would make a good exercise to better understand this material. There are thus three different momentum equations that together comprise the Navier-Stokes Equations.

Momentum Change and Flux

The time rate of change of momentum within the control volume is

$$\frac{\partial}{\partial t}(\rho u)\Delta x \Delta y \Delta z$$

where ∂ / ∂ t is the partial derivative operator with respect to time presented above. The flux of momentum in the x direction into each face of the control volume is the product of the mass flux (mf × ΔA) and the x direction velocity,

$$mf \times u \Delta A$$

where ΔA is the surface area. For the side facing the reader, the momentum flux is thus

$$\rho u u \Delta y \Delta z$$

The momentum flux out the opposite side is

$$-\left(\rho u u + \frac{\partial}{\partial x}(\rho u u)\Delta x\right)\Delta y \Delta z$$

The mass flux into the face with normal vector in the negative y direction is as derived for the continuity equation above, or ρv. The momentum flux into this face is thus

$$\rho v u \Delta x \Delta z$$

For the face opposite, the momentum flux out is

$$-\left(\rho vu + \frac{\partial}{\partial y}(\rho vu)\Delta y\right)\Delta x \Delta z$$

Using a similar analysis, it is easily shown that the momentum flux into and out of the faces with normal vectors in the z direction are

$$\rho wu \Delta x \Delta y$$

and

$$-\left(\rho wu + \frac{\partial}{\partial z}(\rho wu)\Delta z\right)\Delta x \Delta y$$

Our first expression of the momentum equation comes from adding all of these terms together as expressed in the law of the conservation of momentum

$$\frac{\partial(\rho u)}{\partial t}\Delta x \Delta y \Delta z = \rho uu \Delta y \Delta z - \left[\rho uu + \frac{\partial}{\partial x}(\rho uu)\Delta x\right]\Delta y \Delta z$$

$$+\rho vu \Delta x \Delta z - \left[\rho vu + \frac{\partial}{\partial y}(\rho vu)\Delta y\right]\Delta x \Delta z + \rho wu \Delta x \Delta y - \left[\rho wu + \frac{\partial}{\partial z}(\rho wu)\Delta z\right]\Delta x \Delta y + \sum F_x$$

where ΣF_x is the sum of the external forces on the control volume. The momentum fluxes into the control volume cancel with the first part of the momentum fluxes out of the control volume. Performing this cancellation and moving the momentum fluxes to the left hand side of the equation gives

$$\left(\frac{\partial(\rho u)}{\partial t} + \frac{\partial}{\partial x}(\rho uu) + \frac{\partial}{\partial y}(\rho vu) + \frac{\partial}{\partial z}(\rho wu)\right)\Delta x \Delta y \Delta z = \sum F_x$$

Using the product rule, the momentum change and fluxes can be expanded to

$$\left(\rho\frac{\partial u}{\partial t} + \rho u\frac{\partial u}{\partial x} + \rho v\frac{\partial u}{\partial y} + \rho w\frac{\partial u}{\partial z} + u\frac{\partial \rho}{\partial t} + u\frac{\partial(\rho u)}{\partial x} + u\frac{\partial(\rho v)}{\partial y} + u\frac{\partial(\rho w)}{\partial z}\right)\Delta x \Delta y \Delta z = \sum F_x$$

where it is noted that the last four terms in parentheses are the continuity equation times u. Since this must be zero, that leaves.

$$\left(\rho\frac{\partial u}{\partial t}+\rho u\frac{\partial u}{\partial x}+\rho v\frac{\partial u}{\partial y}+\rho w\frac{\partial u}{\partial z}\right)\Delta x\Delta y\Delta z=\sum F_x$$

In the same method we can calculate the force in the y and z directions:

$$\left(\rho\frac{\partial v}{\partial t}+\rho u\frac{\partial v}{\partial x}+\rho v\frac{\partial v}{\partial y}+\rho w\frac{\partial v}{\partial z}\right)\Delta x\Delta y\Delta z=\sum F_y$$

$$\left(\rho\frac{\partial w}{\partial t}+\rho u\frac{\partial w}{\partial x}+\rho v\frac{\partial w}{\partial y}+\rho w\frac{\partial w}{\partial z}\right)\Delta x\Delta y\Delta z=\sum F_z$$

Bernoulli's Equation

Swiss mathematician Daniel Bernoulli (1700–1782) was the first person to study fluid flow mathematically. For his research, Bernoulli imagined a completely nonviscous and incompressible or "ideal" fluid. In this way, he did not have to worry about all the many complications that are present in real examples of fluid flow. The mathematical equations Bernoulli worked out represent only ideal situations, then, but they are useful in many real-life situations.

Bernoulli's equation is one of the most important/useful equations in fluid mechanics. It may be written,

$$\frac{p_1}{pg}+\frac{u_1^2}{2g}+z_1=\frac{p_2}{pg}+\frac{u_2^2}{2g}+z_2$$

We see that from applying equal pressure or zero velocities we get the two equations from the section above. They are both just special cases of Bernoulli's equation.

Bernoulli's equation has some restrictions in its applicability, they are:

1. Flow is steady;
2. Density is constant (which also means the fluid is incompressible);

3. Friction losses are negligible.
4. The equation relates the states at two points along a single streamline, (not conditions on two different streamlines).

All these conditions are impossible to satisfy at any instant in time! Fortunately for many real situations where the conditions are *approximately* satisfied, the equation gives very good results.

The derivation of Bernoulli's Equation:

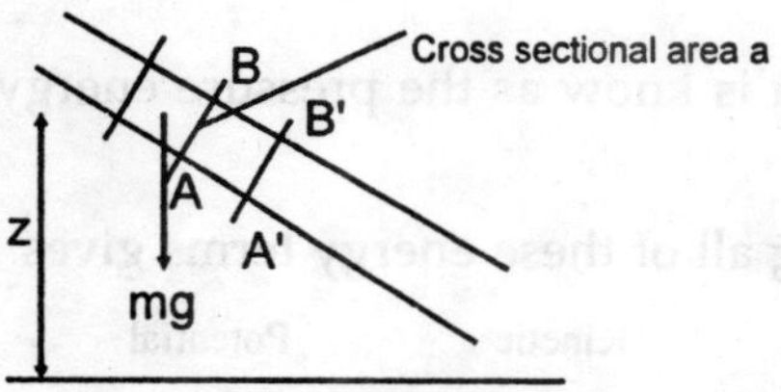

An element of fluid, as that in the figure above, has potential energy due to its height z above a datum and kinetic energy due to its velocity u. If the element has weight mg then

$$\text{potential energy} = mgz$$

$$\text{potential energy per unit weight} = z$$

$$\text{kinetic energy} = \tfrac{1}{2} mu^2$$

$$\text{kinetic energy per unit weight} = u^2/2g$$

At any cross-section the pressure generates a force, the fluid will flow, moving the cross-section, so work will be done. If the pressure at cross section AB is p and the area of the cross-section is a then

$$\text{force on AB} = pa$$

when the mass mg of fluid has passed AB, cross-section AB will have moved to A'B'

$$\text{volume passing AB} = \frac{mg}{pg} = \frac{m}{p}$$

therefore

$$\text{distance } \mathbf{AA'} = \frac{m}{pa}$$

work done = force distance **AA'**

$$= pa \times \frac{m}{pa} = \frac{pm}{p}$$

$$\text{work done per unit weight} = \frac{p}{pg}$$

This term is know as the pressure energy of the flowing stream.

Summing all of these energy terms gives

Pressure		Kinetic		Potential		Total
energy per	+	energy per	+	energy per	=	energy per
unit weight t		unit weight t		unit weight t		unit weight t

or

$$\frac{p}{pg} + \frac{u^2}{2g} + z = H$$

As all of these elements of the equation have units of length, they are often referred to as the following:

$$\text{pressure head} = \frac{p}{pg}$$

$$\text{velocity head} = \frac{u^2}{2g}$$

$$\text{potential head} = z$$

$$\text{total head} = H$$

By the principle of conservation of energy the total *energy* in the system does not change, Thus the total *head* does not change. So the Bernoulli equation can be written.

$$\frac{p}{pg} + \frac{u^2}{2g} + z = H = \text{Constant}$$

As stated above, the Bernoulli equation applies to conditions along a streamline. We can apply it between two points, 1 and 2, on the streamline in the figure below

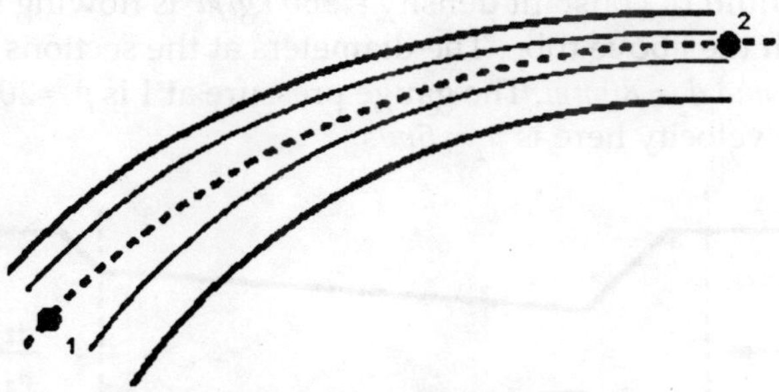

Fig. Two points joined by a streamline

total energy per unit weight at 1=total energy per unit weight at 2

or

total head at 1 = total head at 2

or

$$\frac{p_1}{pg} + \frac{u_1^2}{2g} + z_1 = \frac{p_2}{pg} + \frac{u_2^2}{2g} + z_2$$

This equation assumes no energy losses (e.g. from friction) or energy gains (e.g. from a pump) along the streamline. It can be expanded to include these simply, by adding the appropriate energy terms:

Total	Total	Loss	Work done	Energy
energy per =	energy per unit +	per unit +	per unit –	supplied
unit weight t at 1	weight at 2	weight		weithtper unit w eight

$$\frac{p_1}{pg} + \frac{u_1^2}{2g} + z_1 = \frac{p_2}{pg} + \frac{u_2^2}{2g} + z_2 + h + w - q$$

Example: When the Bernoulli equation is combined with the continuity equation the two can be used to find velocities and pressures at points in the flow connected by a streamline.

Here is an example of using the Bernoulli equation to determine pressure and velocity at within a contracting and expanding pipe.

A fluid of constant density = *960 kg/m³* is flowing steadily through the above tube. The diameters at the sections are d_1 = *100mm and* d_2 = *80mm*. The gauge pressure at 1 is p_1 =*200k N/m²* *and* the velocity here is u_1 = *5m/s.*

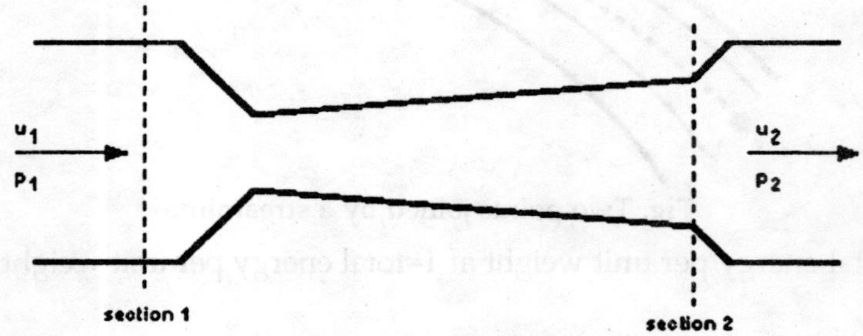

Fig. A contracting expanding pipe

The tube is horizontal, with $z_1 = z_2$ so Bernoulli gives us

$$p_2 = p_1 + \frac{\rho}{2}\left(u_1^2 - u_2^2\right)$$

But we do not know the value of u_2. We can calculate this from the continuity equation: Discharge into the tube is equal to the discharge out i.e.

$$A_1 u_1 = A_2 u_2$$

$$u_2 = \frac{A_1 u_1}{A_2}$$

$$u_2 = \left(\frac{d_1}{d_2}\right)^2 u_1$$

$$= 7.8125\ m/s$$

So

$$
\begin{aligned}
p_2 &= 200000 - 17296.87 \\
&= 182703\ N/m^2 \\
&= 182.7\ kN/m^2
\end{aligned}
$$

Notice how the velocity has increased while the pressure has decreased. The phenomenon - that pressure decreases as velocity increases - sometimes comes in very useful in engineering. (It is on this principle that carburettor in many car engines work - pressure reduces in a contraction allowing a small amount of fuel to enter). Here we have used both the Bernoulli equation and the Continuity principle together to solve the problem.

RAYLEIGH FLOW

Rayleigh flow is (frictionless) flow with heat transfer through a pipe of constant cross sectional area. In practice Rayleigh flow is difficult to achieve. Yet Rayleigh flow is practical and useful concept in a obtaining trends and limits. The density and pressure change due to external cooling or heating. As opposed to the two previous models, the heat transfer can be in two directions not like the friction (there is no negative friction). This fact create situation different compare to the previous two models. This model applied to case where the heat transfer is significant.

The third simple model for an one dimensional flow is for constant heat transfer for frictionless flow. This flow referred in the literature as Rayleigh Flow. This flow is another extreme case in which the friction effect are neglected because their relative effect is much smaller the heat transfer effect. While the isothermal flow model has heat transfer and friction the main assumption was that relative length is so the heat transfer occurs between the surrounding and tube. In contract the heat transfer in Rayleigh flow occurs either between unknown temperature to tube but the heat flux maintained constant. As before, a simple model is build around assumption of constant properties (poorer prediction to case were chemical reaction take palace).

This model usage is to a rough predict the situation involve mostly chemical reaction. In analysis the flow one has to be aware that properties do change significantly for a large range of temperature. Yet, for smaller range of temperature and length the calculations are more accurate. Nevertheless, the main characteristic of the flow such as chocking condition etc. are encapsulated in this model.

The basic physics of the flow revolves around the fact that the gas is highly compressible. The density change though the heat transfer (temperature change). As appose to the Fanno flow in which the resistance always oppose the the flow direction, in Rayleigh flow also cooling can be applied. The flow velocity acceleration change the direction when the cooling is applied.

Chapter 13

The Thermodynamics of Fluids

GRAPHICAL METHODS

Although geometrical representations of propositions in the thermodynamics of fluids are in general use, and have done good service in disseminating clear notions in this science, yet they have by no means received the extension in respect to variety and generality of which they are capable. So far as regards a general graphical method, which can exhibit at once all the thermodynamic properties of a fluid concerned in reversible processes, and serve like for the demonstration of general theorems and the numerical solution of particular problems, it is the general if not the universal practice to use diagrams in which the rectilinear co-ordinates represent volume and pressure. The object of this article is to call attention to certain diagrams of different construction, which afford graphical methods coextensive in their applications with that in ordinary use, and preferable to it in many cases in respect of distinctness or of convenience.

QUANTITIES AND RELATIONS

We have to consider the following quantities:—

v, the volume,

p, the pressure,

t, the (absolute) temperature, of a given body in any state, ε, the energy,

η, the entropy,

also W, the work done, by the body in passing from one state

and H, the heat received, to another.

These are subject to the relations expressed by the following differential equations:—

$$dW = \alpha p dv,$$

$$d\varepsilon = BdH - dW,$$

$$dh = \frac{dH^*}{t}$$

Where α and B are constants depending upon the units by which *v, p, W* and *H* are measured. We may suppose our units so chosen that α = 1 and B = 1, and write our equations in the simpler form,

$d\varepsilon = dH - dW,$

$dW = pdv,$

$dH = td\eta.$

Eliminating *dW* and *dH*, we have

$d\varepsilon = td\eta - pdv.$

The quantities *v, p, t,* ε, and η are determined when the state of the body is given, and it may be permitted to call them *functions of the state of the body*. The state of a body, in the sense in which the term is used in the thermodynamics of fluids, is capable of two independent variations, so that between the five quantities *v, p, t,* ε, and η there exist relations expressible by three finite equations, different in general for different substances, but always such as to be in harmony with the differential equation. This equation evidently signifies that if ε be expressed as function of *v* and η, the partial differential co-efficients of this function taken with respect to *v* and to η will be equal to - *p* and to t respectively.

On the other hand W and H are not functions of the state of the body (or functions of any of the quantities v, p, t, ε, and η), but are determined by the whole series of states through which the body is supposed to pass.

Fundamental Idea and General Properties

Now if we associate a particular point in a plane with every separate state, of which the body is capable, in any continuous manner, so that states differing infinitely little are associated with points which are infinitely near to each other, the points associated with states of equal volume will form lines, which may be called *lines of equal volume*, the different lines being distinguished by the numerical value of the volume (as lines of volume 10, 20, 30, etc.). In the same way we may conceive of *lines of equal pressure, of equal temperature, of equal energy, and of equal entropy*. These lines we may also call *isometric, isopiestic, isothermal, isodynamic, isentropic*, and if necessary use these words as substantives.

Suppose the body to change its state, the points associated with the states through which the body passes will form a line, which we may call the path of the body. The conception of a path must include the idea of direction, to express the order in which the body passes through the series of states. With every such change of state there is connected in general a certain amount of work done, W, and of heat received, H, which we may call the *work* and the *heat* of the *path*. The value of these quantities may be calculated from equations,

$$dW = pdv,$$

$$dH = td\eta,$$

$$\text{i.e., } W = +\int pdv,$$

$$H = +\int td\eta$$

The integration being carried on from the beginning to the end of the path. If the direction of the path is reversed, W and H change their signs, remaining the same in absolute value.

If the changes of state of the body form a cycle, i.e., if the final state is the same as the initial, the path becomes a *circuit*, and the work done and heat received are equal, as may be seen from equation, which when integrated for this case becomes 0 = $H - W$.

The circuit will enclose a certain area, which we may consider as positive or negative according to the direction of the circuit which circumscribes it. The direction in which areas must be circumscribed in order that their value may be positive, is of course arbitrary. In other words, if x and y are the rectangular co-ordinates, we may define an area either as $+\int ydx$, or as $+\int xdy$.

If an area be divided into any number of parts, the work done in the circuit bounding the whole area is equal to the sum of the work done in all the circuits bounding the partial areas. This is evident from the consideration, that the work done in each of the lines which separate the partial areas appears twice and with contrary signs in the sum of the work done in the circuits bounding the partial areas. Also the heat received in the circuit bounding the whole area is equal to the sum of the heat received in all the circuits bounding the partial areas.

If all the dimensions of a circuit are infinitely small, the ratio of the included area to the work or heat of the circuit is independent of the shape of the circuit and the direction in which it is described, and varies only with its position in the diagram.

That this ratio is independent of the direction in which the circuit is described, is evident from the consideration that a reversal of this direction simply changes the sign of both terms of the ratio. To prove that the ratio is independent of the shape of the circuit, let us suppose the area ABCDE divided up by an infinite number of isometrics $v_1 v_1$, $v_2 v_2$, etc., with equal differences of volume dv, and an infinite number of isopiestics $p_1 p_1$, $p_2 p_2$, etc., with equal differences of pressure dp. Now from the principle of continuity, as the whole figure is infinitely small, the ratio of the area of one of the small quadrilaterals into which the figure is divided to the work done in passing around it is approximately the same for all the different quadrilaterals.

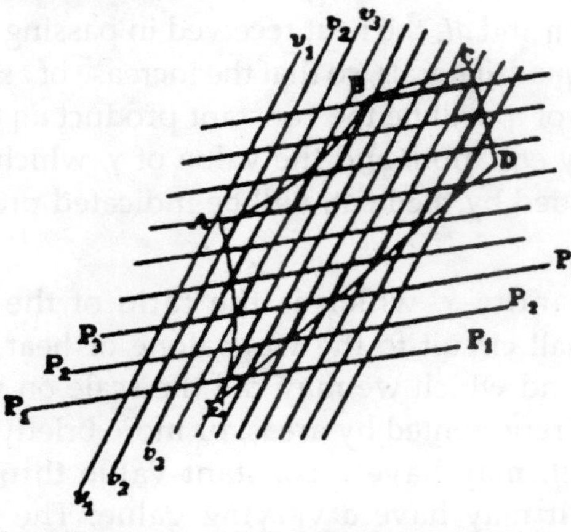

Therefore the area of the figure composed of all the complete quadrilaterals which fall within the given circuit has to the work done in circumscribing this figure the same ratio, which we will call γ. But the area of this figure is approximately the same as that of the given circuit, and the work done in describing this figure is approximately the same as that done in describing the given circuit. Therefore the area of the given circuit has to the work done or heat received in that circuit this ratio γ, which is independent of the shape of the circuit.

Now if we imagine the systems of equidifferent isometrics and isopiestics, which have just been spoken of, extended over the whole diagram, the work done in circumscribing one of the small quadrilaterals, so that the increase of pressure directly precedes the increase of volume, will have in every part of the diagram a constant value, viz, the product of the differences of volume and pressure ($dv \times dp$), as may easily be proved by applying equation successively to its four sides. But the area of one of these quadrilaterals, which we could consider as constant within the limits of the infinitely small circuit, may vary for different parts of the diagram, and will indicate proportionally the value of γ, which is equal to the area divided by $dv \times dp$.

In like manner, if we imagine systems of isentropics and isothermals drawn throughout the diagram for equal differences $d\eta$ and dt, the heat received in passing around one of the small quadrilaterals, so that the increase of t shall directly precede that of η, will be the constant product $d\eta \times dt$, as may be proved by equation, and the value of γ, which is equal to the area divided by the heat, will be indicated proportionally by the areas.

This quantity γ, which is the ratio of the area of an infinitely small circuit to the work done or heat received in that circuit, and which we may call the scale on which work and heat are represented by areas, or more briefly, the *scale of work and heat*, may have a constant value throughout the diagram or it may have a varying value. The diagram in ordinary use affords an example of the first case, as the area of a circuit is everywhere proportional to the work or heat. There are other diagrams which have the same property, and we may call all such *diagrams of constant scale*.

In any case we may consider the scale of work and heat as known for every point of the diagram so far as we are able to draw the isometrics and isopiestics or the isentropics and isothermals. If we write δW and δH for the work and heat of an infinitesimal circuit, and δA for the area included, the relations of these quantities are thus expressed:—

$$\delta W = \delta H = \frac{1}{\gamma}\delta A.$$

We may find the value of W and H for a circuit of finite dimensions by supposing the included area A divided into areas δA infinitely small in all directions, for which therefore the above equation will hold, and taking the sum of the values of δH or δW for the various areas δA. Writing W^c and H^c for the work and heat of the circuit C, and Σ^c for a summation or integration performed within the limits of this circuit, we have

$$W^c = H^c = \sum c \frac{1}{\gamma} \delta A$$

We have thus an expression for the value of the work and heat of a circuit involving an integration extending over an area instead of one extending over a line, as in equations.

Similar expressions may be found for the work and the heat of a path which is not a circuit. For this case may be reduced to the preceding by the consideration that $W = 0$ for a path on an isometric or on the line of no pressure, and $H = 0$ for a path on an isentropic or on the line of absolute cold. Hence the work of any path S is equal to that of the circuit formed of S, the isometric of the final state, the line of no pressure and the isometric of the initial state, which circuit may be represented by the notation $[S, v'', p^0, v']$. And the heat of the same path is the same as that of the circuit $[S, \eta'', t^0, \eta']$. Therefore using W^s and H^s to denote the work and heat of any path S, we have

$$W^s = \sum^{[S, V'', p^0, v']} \frac{1}{\gamma} \delta A,$$

$$H^s = \sum^{[s, n'', t^0, n']} \frac{1}{\gamma} \delta A$$

Where as before the limits of the integration are denoted by the expression occupying the place of an index to the sign Σ. These equations evidently include equation as a particular case.

It is easy to form a material conception of these relations. If we imagine, for example, mass inherent in the plane of the diagram with a varying (superficial) density represented by

$$\frac{1}{\gamma}$$

, then

$$\sum \frac{1}{\gamma} \delta A$$

Will evidently denote the mass of the part of the plane included within the limits of integration, this mass being taken positively or negatively according to the direction of the circuit.

Thus far we have made no supposition in regard to the nature of the law, by which we associate the points of a plane with the states of the body, except a certain condition of continuity.

Whatever law we may adopt, we obtain a method of representation of the thermodynamic properties of the body, in which the relations existing between the functions of the state of the body are indicated by a net-work of lines, while the work done and the heat received by the body when it changes its state are represented by integrals extending over the elements of a line, and also by an integral extending over the elements of certain areas in the diagram, or, if we choose to introduce such a consideration, by the mass belonging to these areas.

The different diagrams which we obtain by different laws of association are all such as may be obtained from one another by a process of *deformation,* and this consideration is sufficient to demonstrate their properties from the well-known properties of the diagram in which the volume and pressure are represented by rectangular coordinates.

For the relations indicated by the net-work of isometrics, isopiestics etc are evidently not altered by deformation of the surface upon which they are drawn, and if we conceive of mass as belonging to the surface, the mass included within given lines will also not be affected by the process of deformation. If, then, the surface upon which the ordinary diagram is drawn has the uniform superficial density 1, so that the work and heat of a circuit, which are represented in this diagram by the included area, shall also be represented by the mass included, this latter relation will hold for any diagram formed from this by deformation of the surface on which it is drawn.

The choice of the method of representation is of course to be determined by considerations of simplicity and convenience, especially in regard to the drawing of the lines of equal volume, pressure, temperature, energy and entropy, and the estimation of work and heat. There is an obvious advantage in the use of diagrams of constant scale, in which the work and heat are represented simply by areas.

Such diagrams may of course be produced by an infinity of different methods, as there is no limit to the ways of deforming a plane figure without altering the magnitude of its elements. Among these methods, two are especially important,—the ordinary method in which the volume and pressure are represented by rectilinear co-ordinates, and that in which the entropy and temperature are so represented. A diagram formed by the former method may be called, for the sake of distinction, a *volume-pressure* diagram,—one formed by the latter, an *entropytemperature* diagram. That the latter as well as the former satisfies the condition that $\gamma = 1$ throughout the whole diagram.

The Entropy-temperature

As the general equations are not altered by interchanging v, $-p$ and $-W$ with η, t and H respectively, it is evident that, so far as these equations are concerned, there is nothing to choose between a volume-pressure and an entropy-temperature diagram. In the former, the work is represented by an area bounded by the path which represents the change of state of the body, two ordinates and the axis of abscissas. The same is true of the heat received in the latter diagram.

Again, in the former diagram, the heat received is represented by an area bounded by the path and certain lines, the character of which depends upon the nature of the body under consideration. Except in the case of an ideal body, the properties of which are determined by assumption, these lines are more or less unknown in a part of their course, and in any case the area will generally extend to an infinite distance. Very much the same inconveniences attach themselves to the areas representing work in the entropy temperature diagram.

There is, however, a consideration of a general character, which shows an important advantage on the side of the entropy-temperature diagram. In thermodynamic problems, heat received at one temperature is by no means the equivalent of the same amount of heat received at another temperature. For example, a supply of a million calories at 150^c is a very different thing from a supply of a million calories at 50^c. But no such distinction exists in regard to work. This is a result of the general law, that heat can only pass from a hotter to a colder body, while work can be transferred by mechanical means from one fluid to any other, whatever may be the pressures.

Hence, in thermodynamic problems, it is generally necessary to distinguish between the quantities of heat received or given out by the body at different temperatures, while as far as work is concerned, it is generally sufficient to ascertain the total amount performed. If, then, several heat-areas and one work-area enter into the problem, it is evidently more important that the former should be simple in form, than that the latter should be so. Moreover, in the very common case of a circuit, the work-area is bounded entirely by the path, and the form of the isometrics and the line of no pressure are of no especial consequence.

It is worthy of notice that the simplest form of a perfect thermodynamic engine, so often described in treatises on thermodynamics, is represented in the entropy-temperature diagram by a figure of extreme simplicity, viz: a rectangle of which the sides are parallel to the co-ordinate axes.

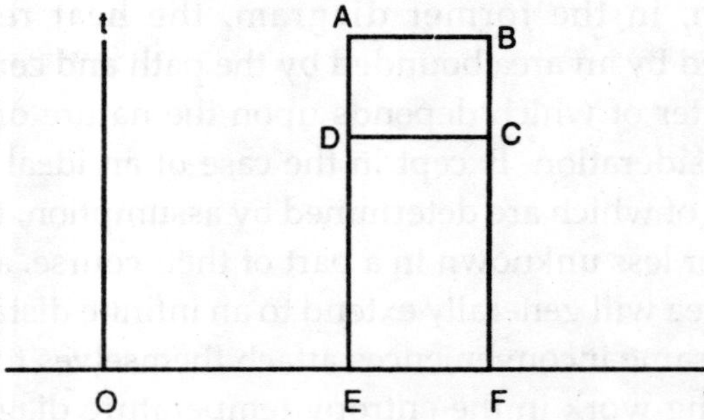

Thus, the circuit ABCD may represent the series of states through which the fluid is made to pass in such an engine, the included area representing the work done, while the area ABFE represents the heat received from the heater at the highest temperature AE, and the area CDEF represents the heat transmitted to the cooler at the lowest temperature DE.

There is another form of the perfect thermodynamic engine, viz: one with a perfect regenerator as defined by Rankine, the representation of which becomes peculiarly simple in the entropy-temperature diagram. The circuit consists of two equal straight lines AB and CD parallel to the axis of abscissas, and two precisely similar curves of any form BC and AD.

The included area ABCD represents the work done, and the areas ABba and CDdc represent respectively the heat received from the heater and that transmitted to the cooler. The heat imparted by the fluid to the regenerator in passing from B to C, and afterward restored to the fluid in its passage from D to A, is represented by the areas BCcb and DAad.

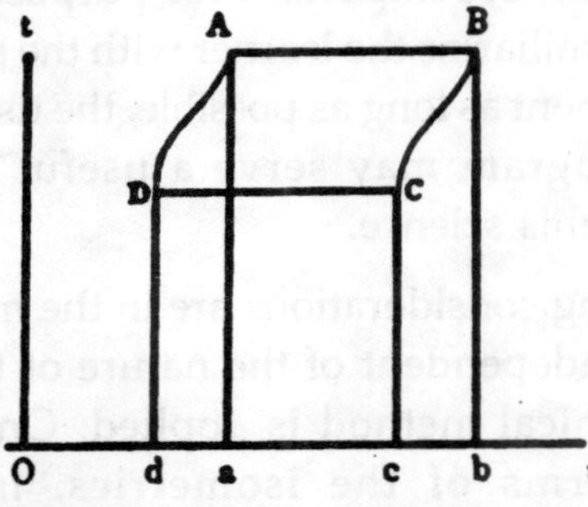

Fig. It is often a matter of the first importance in the study of any thermodynamic engine, to compare it with a perfect engine. Such a comparison will obviously be much facilitated by the use of a method in which the perfect engine is represented by such simple forms.

The method in which the co-ordinates represent volume and pressure has a certain advantage in the simple and elementary character of the notions upon which it is based, and its analogy with Watt's indicator has doubtless contributed to render it popular. On the other hand, a method involving the notion of *entropy*, the very existence of which depends upon the second law of thermodynamics, will doubtless seem to many far-fetched, and may repel beginners as obscure and difficult of comprehension.

This inconvenience is perhaps more than counterbalanced by the advantages of a method which makes the second law of thermodynamics so prominent, and gives it so clear and elementary an expression. The fact, that the different states of a fluid can be represented by the positions of a point in a plane, so that the ordinates shall represent the temperatures, and the heat received or given out by the fluid shall be represented by the area bounded by the line representing the states through which the body passes, the ordinates drawn through the extreme points of this line, and the axis of abscissas,—this fact, clumsy as its expression in words may be, is one which presents a clear image to the eye, and which the mind can readily grasp and retain.

If, then, it is more important for purposes of instruction and the like to familiarize the learner with the second law, than to defer its statement as long as possible, the use of the entropy-temperature diagram may serve a useful purpose in the popularizing of this science.

The foregoing considerations are in the main of a general character, and independent of the nature of the substance to which the graphical method is applied. On this, however, depend the forms of the isometrics, isopiestics and isodynamics in the entropy-temperature diagram, and of the isentropics, isothermals and isodynamics in the volume-pressure diagram. As the convenience of a method depends largely upon the ease with which these lines can be drawn, and upon the peculiarities of the fluid which has its properties represented in the diagram, it is desirable to compare the methods under consideration in some of their most important applications. We will commence with the case of a perfect gas.

Case of a Perfect Gas

A perfect or ideal gas may be defined as such a gas, that for any constant quantity of it the product of the volume and the pressure varies as the temperature, and the energy varies as the temperature, i.e.,

$pv = at,$

$\varepsilon = ct.$

The significance of the constant a is sufficiently indicated by equation. The significance of c may be rendered more evident by differentiating equation and comparing the result

$d\varepsilon = cdt$

With the general equations and, viz:

$d\varepsilon = dH - dW, dW = pdv.$

If $dv = 0$, $dW = 0$, and $dH = cdt$, i.e.

$$\left(\frac{dH}{dt}\right)_v = c,$$

i.e., c is the quantity of heat necessary to raise the temperature of the body one degree under the condition of constant volume. It will be observed, that when different quantities of the same gas are considered, a and c both vary as the quantity, and $c \chi a$ is constant; also, that the value of $c \chi a$ for different gases varies as their specific heat determined for equal volumes and for constant volume.

Any other equations which subsist between v, p, t, ε and η may be derived from the three independent equations (A), (B) and (D). If we eliminate ε from (B) and (D), we have

$\eta = a \log v + c \log t + c \log c.$

Eliminating v from equation, we have

$\eta = (a + c) \log t - a \log p + c \log c + a \log a.$

Eliminating t, we have

$$\eta = (\alpha + c)\log v + c\log p + c\log\frac{c}{a}.$$

If v is constant, equation becomes

$\eta = c \log t + \text{Const.},$

i.e., the isometrics in the entropy-temperature diagram are logarithmic curves identical with one another in form,—a change in the value of v having only the effect of moving the curve parallel to the axis of η. If p is constant, equation (F) becomes

$\eta = (a + c) \log t + \text{Const.},$

So that the isopiestics in this diagram have similar properties. This identity in form diminishes greatly the labour of drawing any considerable number of these curves. For if a card or thin board be cut in the form of one of them, it may be used as a pattern or ruler to draw all of the same system.

The isodynamics are straight in this diagram (eq B).

To find the form of the isothermals and isentropics in the volume pressure diagram, we may make t and η constant in equations (A) and (G) respectively, which will then reduce to the well-known equations of these curves:—

$pv = \text{Const.},$

And $p^c v^{a+c} = \text{Const.}$

The equation of the isodynamics is of course the same as that of the isothermals. None of these systems of lines have that property of identity of form, which makes the systems of isometrics and isopiestics so easy to draw in the entropy-temperature diagram.

Case of Condensable Vapours

The case of bodies which pass from the liquid to the gaseous condition is next to be considered. It is usual to assume of such a body, that when sufficiently superheated it approaches the condition of a perfect gas. If, then, in the entropy-temperature diagram of such a body we draw systems of isometrics, isopiestics and isodynamics, as if for a perfect gas, for proper values of the constants a and c, these will be asymptotes to the true isometrics, etc., of the vapour, and in many cases will not vary from them greatly in the part of the diagram which represents vapour unmixed with liquid, except in the vicinity of the line of saturation. In the volume-pressure diagram of the same body, the isothermals, isentropics and isodynamics, drawn for a perfect gas for the same values of a and c, will have the same relations to the true isothermals, etc.

In that part of any diagram which represents a mixture of vapour and liquid, the isopiestics and isothermals will be identical, as the pressure is determined by the temperature alone. In both the diagrams which we are now comparing, they will be straight and parallel to the axis of abscissas. The form of the isometrics and isodynamics in the entropy-temperature diagram, or that of the isentropics and isodynamics in the volume-pressure diagram, will depend upon the nature of the fluid, and probably cannot be expressed by any simple equations. The following property, however, renders it easy to construct equidifferent systems of these lines, viz: any such system will divide any isothermal (isopiestic) into equal segments.

It remains to consider that part of the diagram which represents the body when entirely in the condition of liquid. The fundamental characteristic of this condition of matter is that the volume is very nearly constant, so that variations of volume are generally entirely inappreciable when represented graphically on the same scale on which the volume of the body in the state of vapour is represented, and both the variations of volume and the connected variations of the connected quantities may be, and generally are, neglected by the side of the variations of the same quantities which occur when the body passes to the state of vapour.

Let us make, then, the usual assumption that v is constant, and see how the general equations,are thereby affected.

We have first,

$dv = 0$,

Then $dW = 0$,

And $d\varepsilon = t d\eta$.

If we add $dH = td\eta$,

These four equations will evidently be equivalent to the three independent equations, combined with the assumption which we have just made. For a liquid, then, ε, instead of being a function of two quantities v and η, is a function of η alone,—t is also a function of η alone, being equal to the differential co-efficient of the function ε; that is, the value of one of the three quantities t, ε and η, is sufficient to determine the other two. The value of v, moreover, is fixed without reference to the values of t, ε and η (so long as these do not pass the limits of values possible for liquidity); while p does not enter into the equations, i.e., p may have any value (within certain limits) without affecting the values of t, ε, η or v.

If the body change its state, continuing always liquid, the value of W for such a change is 0, and that of H is determined by the values of any one of the three quantities t, ε and η. It is, therefore, the relations between t, ε, η and H, for which a graphical expression is to be sought; a method, therefore, in which the co-ordinates of the diagram are made equal to the volume and pressure, is totally inapplicable to this particular case; v and p are indeed the only two of the five functions of the state of the body, v, p, t, ε, and η, which have no relations either to each other, or to the other three, or to the quantities W and H, to be expressed.

The values of v and p do not really determine the state of an incompressible fluid,—the values of t, ε, and η are still left undetermined, so that through every point in the volume-pressure diagram which represents the liquid there must pass (in general) an infinite number of isothermals, isodynamics and isentropics. The character of this part of the diagram is as follows:—the states of liquidity are represented by the points of a line parallel to the axis of pressures, and the isothermals, isodynamics and isentropics, which cross the field of partial vapourization and meet this line, turn upward and follow its course.

The line of liquidity is a curve AB determined by the relation between t and η. This curve is also an isometric. Every point of it has a definite volume, temperature, entropy and energy. The latter is indicated by the isodynamics $E_1 E_1$, $E_2 E_2$, etc., which cross the region of partial vapourization and terminate in the line of liquidity. (They do not in this diagram turn and follow the line.)

If the body pass from one state to another, remaining liquid, as from M to N in the figure, the heat received is represented as usual by the area MNnm That the work done is nothing, is indicated by the fact that the line AB is an isometric. Only the isopiestics in this diagram are superposed in the line of fluidity, turning downward where they meet this line and following its course, so that for any point in this line the pressure is undetermined. This is, however, no inconvenience in the diagram, as it simply expresses the fact of the case, that when all the quantities v, t, ε, and η are fixed, the pressure is still undetermined.

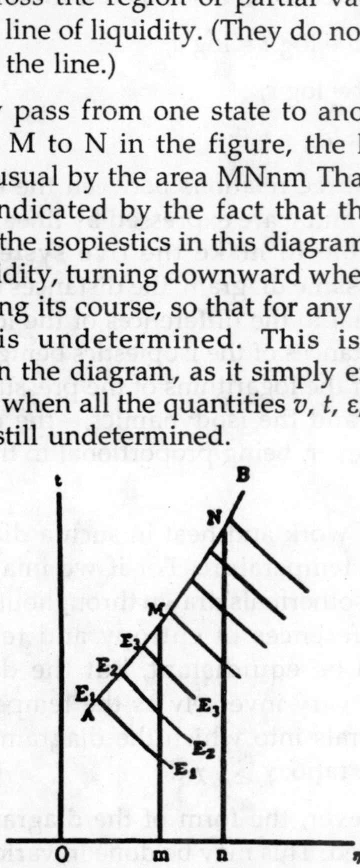

Fig. Diagram in which the Isometrics, Isopiestics, Isothermals, Isodynamics and Isentropics of a Perfect Gas are all Straight Lines

There are many cases in which it is of more importance that it should be easy to draw the lines of equal volume, pressure, temperature, energy and entropy, than that work and heat should be represented in the simplest manner. In such cases it may be expedient to give up the condition that the scale (γ) of work and heat shall be constant, when by that means it is possible to gain greater simplicity in the form of the lines just mentioned.

In the case of a perfect gas, the three relations between the quantities v, p, t, ε, and η are given, equations. These equations may be easily transformed into the three

$$\text{Log } p + \log v - \log t = \log a,$$

$$\text{Log } \varepsilon - \log t = \log c,$$

$$\eta - c \log \varepsilon - a \log v = 0;$$

so that the three relations between the quantities log v, log p, log t, log ε and η are expressed by linear equations, and it will be possible to make the five systems of lines all rectilinear in the same diagram, the distances of the isometrics being proportional to the differences of the logarithms of the volumes, the distances of the isopiestics being proportional to the differences of the logarithms of the pressures, and so with the isothermals and the isodynamics,—the distances of the isentropics, however, being proportional to the differences of entropy simply.

The scale of work and heat in such a diagram will vary inversely as the temperature. For if we imagine systems of isentropics and isothermals drawn throughout the diagram for equal small differences of entropy and temperature, the isentropics will be equidistant, but the distances of the isothermals will vary inversely as the temperature, and the small quadrilaterals into which the diagram is divided will vary in the same ratio: $\gamma \geq 1 \chi$ t.

So far, however, the form of the diagram has not been completely defined. This may be done in various ways: e.g., if x and y be the rectangular co-ordinates, we may make

$$x = \log v,$$

$$y = \log p;$$

or $x = \eta, y = \log t;$

or $x = \log v, y = \eta;$ etc.

Or we may set the condition that the logarithms of volume, of pressure and of temperature, shall be represented in the diagram on the same scale. (The logarithms of energy are necessarily represented on the same scale as those of temperature.) This will require that the isometrics, isopiestics and isothermals cut one another at angles of 60°.

The general character of all these diagrams, which may be derived from one another by projection by parallel lines, may be illustrated by the case in which $x = \log v$, and $y = \log p$.

Through any point A of such a diagram let there be drawn the isometric vv′, the isopiestic pp′, the isothermal tt′, and the isentropic ηη′. The lines pp′ and vv′ are of course parallel to the axes.

$$\tan tAp = \left(\frac{dy}{dx}\right)_t = \left(\frac{d\log p}{d\log v}\right)_t = -1$$

The Volume-entropy Diagram

The method of representation, in which the co-ordinates of the point in the diagram are made equal to the volume and entropy of the body, presents certain characteristics which entitle it to a somewhat detailed consideration, and for some purposes give it substantial advantages over any other method. We might anticipate some of these advantages from the simple and symmetrical form of the general equations of thermodynamics, when volume and entropy are chosen as independent variables, viz:—

$$p = -\frac{d\in}{dv}, t = \frac{d\in}{d\eta}$$

$$dW = pdv,$$

$$dH = tdn.$$

Eliminating p and t we have also

$$dW = -\frac{d\in}{dv}dv, dH = \frac{d\in}{d\eta}d\eta,$$

The geometrical relations corresponding to these equations are in the volume-entropy diagram extremely simple. To fix our ideas, let the axes of volume and entropy be horizontal and vertical respectively, volume increasing toward the right and entropy upward. Then the pressure taken negatively will equal the ratio of the difference of energy to the difference of volume of two adjacent points in the same horizontal line, and the temperature will equal the ratio of the difference of energy to the difference of entropy of two adjacent points in the same vertical line. Or, if a series of isodynamics be drawn for equal infinitesimal differences of energy, any series of horizontal lines will be divided into segments inversely proportional to the pressure, and any series of vertical lines into segments inversely proportional to the temperature.

We see by equations and, that for a motion parallel to the axis of volume, the heat received is 0, and the work done is equal to the decrease of the energy, while for a motion parallel to the axis of entropy, the work done is 0, and the heat received is equal to the increase of the energy. These two propositions are true either for elementary paths or for those of finite length. In general, the work for any element of a path is equal to the product of the pressure in that part of the diagram into the horizontal projection of the element of the path, and the heat received is equal to the product of the temperature into the vertical projection of the element of the path.

If we wish to estimate the value of the integrals $+\int p dv$ and $+\int t d\eta$, which represent the work and heat of any path, by means of measurements upon the diagram, or if we wish to appreciate readily by the eye the approximate value of these expressions, or if we merely wish to illustrate their meaning by means of the diagram; for any of these purposes the diagram which we are now considering will have the advantage that it represents the differentials dv and $d\eta$ more simply and clearly than any other.

But we may also estimate the work and heat of any path by means of an integration extending over the elements of an area, viz: by the formula,

$$W^c = H^c = \sum{}^c \frac{1}{\gamma}\delta A$$

$$W^s = \sum^{\left[s,v'',p^0,v'\right]} \frac{1}{\gamma}\delta A$$

$$H^s = \sum^{\left[s,n'',t^0,n'\right]} \frac{1}{\gamma}\delta A$$

In regard to the limits of integration in these formula, we see that for the work of any path which is not a circuit, the bounding line is composed of the path, the line of no pressure and two vertical lines, and for the heat of the path, the bounding line is composed of the path, the line of absolute cold and two horizontal lines.

As the sign of γ, as well as that of δA, will be indeterminate until we decide in which direction an area must be circumscribed in order to be considered positive, we will call an area positive which is circumscribed in the direction in which the hands of a watch move. This choice, with the positions of the axes of volume and entropy which we have supposed, will make the value of γ in most cases positive

The value of γ, in a diagram drawn according to this method, will depend upon the properties of the body for which the diagram is drawn. In this respect, this method differs from all the others which have been discussed in detail in this article. It is easy to find an expression for γ depending simply upon the variations of the energy, by comparing the area and the work or heat of an infinitely small circuit in the form of a rectangle having its sides parallel to the two axes.

Let $N_1 N_2 N_3 N_4$ be such a circuit, and let it be described in the order of the numerals, so that the area is positive. Also let ε_1, ε_2, ε_3, ε_4 represent the energy at the four corners. The work done in the four sides in order commencing at N_1, will be $\varepsilon_1 - \varepsilon_2$, 0, $\varepsilon_3 - \varepsilon_4$, 0. The total work, therefore, for the rectangular circuit is

$\varepsilon_1 - \varepsilon_2 + \varepsilon_3 - \varepsilon_4$.

Now as the rectangle is infinitely small, if we call its sides *dv* and *dη*, the above expression will be equivalent to

$$-\frac{d^2 \in}{dvd\eta} dvd\eta$$

Dividing by the area *dv dη*, and writing *γv,η* for the scale of work and heat in a diagram of this kind, we have

$$\frac{1}{\gamma_{v,\eta}} = -\frac{d^2 \in}{dvd\eta} = \frac{dp}{d\eta} = -\frac{dt}{dv}$$

The two last expressions for the value of 1 χ *γv,η* indicate that the value of *γv,η* in different parts of the diagram will be indicated proportionally by the segments into which vertical lines are divided by a system of equidifferent isopiestics, and also by the segments into which horizontal lines are divided by a system of equidifferent isothermals.

As, in almost all cases, the pressure of a body is increased when it receives heat without change of volume, $\frac{dp}{d\eta}$ is in general positive, and the same will be true of *γv,η* under the assumptions which we have made in regard to the directions of the axes and the definition of a positive area.

In the estimation of work and heat it may often be of use to consider the deformation necessary to reduce the diagram to one of constant scale for work and heat. Now if the diagram be so deformed that each point remains in the same vertical line, but moves in this line so that all isopiestics become straight and horizontal lines at distances proportional to their differences of pressure, it will evidently become a volume-pressure diagram. Again, if the diagram be so deformed that each point remains in the same horizontal line, but moves in it so that isothermals become straight and vertical lines at distances proportional to their differences of temperature, it will become an entropy-temperature diagram. These considerations will enable us to compute numerically the work or heat of any path which is given in a volume-entropy diagram, when the pressure and temperature are known for all points of the path, in a manner analogous to that explained.

The ratio of any element of area in the volume-pressure or the entropy-temperature diagram, or in any other in which the scale of work and heat is unity, to the corresponding element in the volume entropy diagram is represented by $\frac{1}{\gamma_{v,n}}$ or $-\frac{d^2 \in}{dvd\eta}$ The cases in which this ratio is 0, or changes its sign, demand especial attention, as in such cases the diagrams of constant scale fail to give a satisfactory representation of the properties of the body, while no difficulty or inconvenience arises in the use of the volume-entropy diagram.

As, $-\frac{d^2 \in}{dvd\eta} = \frac{dp}{d\eta}$ value is evidently zero in that part of the diagram which represents the body when in part solid, in part liquid, and in part vapour. The properties of such a mixture are very simply and clearly exhibited in the volume-entropy diagram.

Let the temperature and the pressure of the mixture, which are independent of the proportions of vapour, solid and liquid, be denoted by t' and p'. Also let V, L and S be points of the diagram which indicate the volume and entropy of the body in three perfectly defined states, viz: that of a vapour of temperature $t2$ and pressure $p2$, that of a liquid of the same temperature and pressure, and that of a solid of the same temperature and pressure. And let $v_v, \eta_v, v_L, \eta_L, v_s, \eta_s$ denote the volume and entropy of these states. The position of the point which represents the body, when part is vapour, part liquid, and part solid, these parts being as μ, v, and $1 - \mu - v$, is determined by the equations

$$v = \mu_v + vv_L + (1 - \mu - v)v_s,$$

$$\eta = \mu\eta_v + v\eta_L + (1 - \mu - v)\eta_s,$$

Where v and η are the volume and entropy of the mixture. The truth of the first equation is evident. The second may be written

$\eta - \eta_s = \mu(\eta_v - \eta_s) + v(\eta_L - \eta_s)$,

Or multiplying by $t2$,

$t'(\eta - \eta s) = \mu t'(\eta_v - \eta_s) + vt'(\eta_L - \eta_s)$.

The first member of this equation denotes the heat necessary to bring the body from the state S to the state of the mixture in question under the constant temperature t', while the terms of the second member denote separately the heat necessary to vapourize the part μ, and to liquefy the part v of the body.

The values of v and η are such as would give the centre of gravity of masses μ, v and $1 - \mu - v$ placed at the points V, L and S. Hence the part of the diagram which represents a mixture of vapour, liquid and solid, is the triangle VLS. The pressure and temperature are constant for this triangle, i.e., an isopiestic and also an isothermal here expand to cover a space. The isodynamics are straight and equidistant for equal differences of energy. For $\frac{d\in}{dv} = -p'$ and $\frac{d\in}{d\eta} = t'$, both of which are constant throughout the triangle.

This case can be but very imperfectly represented in the volume pressure, or in the entropy-temperature diagram. For all points in the same vertical line in the triangle VLS will, in the volume-pressure diagram, be represented by a single point, as having the same volume and pressure. And all the points in the same horizontal line will be represented in the entropy-temperature diagram by a single point, as having the same entropy and temperature. In either diagram, the whole triangle reduces to a straight line. It must reduce to a line in any diagram whatever of constant scale, as its area must become 0 in such a diagram.

This must be regarded as a defect in these diagrams, as essentially different states are represented by the same point. In consequence, any circuit within the triangle VLS will be represented in any diagram of constant scale by two paths of opposite directions superposed, the appearance being as if a body should change its state and then return to its original state by inverse processes, so as to repass through the same series of states. It is true that the circuit in question is like this combination of processes in one important particular, viz: that $W = H = 0$, i.e., there is no transformation of heat into work. But this very fact, that a circuit without transformation of heat into work is possible, is worthy of distinct representation.

A body may have such properties that in one part of the volume entropy diagram $\frac{1}{\gamma_{v,\eta}}$ i.e., $\frac{dp}{d\eta}$ is positive and in another negative. These parts of the diagram may be separated by a line, in which $\frac{dp}{d\eta} = 0$, or by one in which $\frac{dp}{d\eta}$ changes abruptly from a positive to a negative value. (In part, also, they may be separated by an area in which $\frac{dp}{d\eta} = 0$.) In the representation of such cases in any diagram of constant scale, we meet with a difficulty of the following nature.

Let suppose that on the right of the line LL in a volume entropy diagram, $\frac{dp}{d\eta}$ is positive, and on the left negative. Then, if we draw any circuit ABCD on the right side of LL, the direction being that of the hands of a watch, the work and heat of the circuit will be positive. But if we draw any circuit EFGH in the same direction on the other side of the line LL, the work and heat will be negative. For

$$W = H = \sum \frac{1}{\gamma_{v,\eta}} \delta A = \sum \frac{dp}{d\eta} \delta A$$

And the direction of the circuits makes the areas positive in both cases. Now if we should change this diagram into any diagram of constant scale, the areas of the circuits, as representing proportionally the work done in each case, must necessarily have opposite signs, ie., the direction of the circuits must be opposite. We will suppose that the work done is positive in the diagram of constant scale, when the direction of the circuit is that of the hands of a watch. Then, in that diagram, the circuit ABCD would have that direction and the circuit EFGH the contrary direction.

Now if we imagine an indefinite number of circuits on each side of LL in the volume-entropy diagram, it will be evident that to transform such a diagram into one of constant scale, so as to change the direction of all the circuits on one side of LL, and of none on the other the diagram must be *folded over* along that line; so that the points on one side of LL in a diagram of constant scale do not represent any states of the body, while on the other side of this line, each point, for a certain distance at least, represents two different states of the body, which in the volume-entropy diagram are represented by points on opposite sides of the line LL.

We have thus in a part of the field two diagrams superposed, which must be carefully distinguished. If this be done, as by the help of different colours, or of continuous and dotted lines, or otherwise, and it is remembered that there is no continuity between these superposed diagrams, except along the bounding line LL, all the general theorems which have been developed in this article can be readily applied to the diagram. But to the eye or to the imagination, the figure will necessarily be much more confusing than a volume-entropy diagram.

If

$$\frac{dp}{d\eta}=0$$

for the line LL, there will be another inconvenience in the use of any diagram of constant scale, viz: in the vicinity of the line LL,

$\frac{dp}{d\eta}$, i.e., $1\ \chi\ \gamma_{v'}\eta$ will have a very small value, so that areas will be very greatly reduced in the diagram of constant scale, as compared with the corresponding areas in the volume-entropy diagram. Therefore, in the former diagram, either the isometrics, or the isentropics, or both, will be crowded together in the vicinity of the line LL, so that this part of the diagram will be necessarily indistinct.

It may occur, however, in the volume-entropy diagram, that the same point must represent two different states of the body. This occurs in the case of liquids which can be vapourized. Let MM be the line representing the states of the liquid bordering upon vapourization. This line will be near to the axis of entropy, and nearly parallel to it. If the body is in a state represented by a point of the line MM, and is compressed without addition or subtraction of heat, it will remain of course liquid. Hence, the points of the space immediately on the left of MM represent simple liquid.

On the other hand, the body being in the original state, if its volume should be increased without addition or subtraction of heat, and if the conditions necessary for vapourization are present (conditions relative to the body enclosing the liquid in question, etc.), the liquid will become partially vapourized, but if these conditions are not present it will continue liquid.

Hence, every point on the right of MM and sufficiently near to it represents two different states of the body in one of which it is partially vapourized, and in the other it is entirely liquid. If we take the points as representing the mixture of vapour and liquid, they form one diagram, and if we take them as representing simple liquid, they form a totally different diagram superposed on the first. There is evidently no continuity between these diagrams except at the line MM; we may regard them as upon separate sheets united only along MM. For the body cannot pass from the state of partial vapourization to the state of liquid except at this line.

The reverse process is indeed possible; the body can pass from the state of superheated liquid to that of partial vapourization, if the conditions of vapourization alluded to above are supplied, or if the increase of volume is carried beyond a certain limit, but not by gradual changes or reversible processes.

After such a change, the point representing the state of the body will be found in a different position from that which it occupied before, but the change of state cannot be properly represented by any path, as during the change the body does not satisfy that condition of uniform temperature and pressure which has been assumed throughout this article, and which is necessary for the graphical methods under discussion.

Of the two superposed diagrams, that which represents simple liquid is a continuation of the diagram on the left of MM. The isopiestics, isothermals and isodynamics pass from one to the other without abrupt change of direction or curvature. But that which represents a mixture of vapour and liquid will be different in its character, and its isopiestics and isothermals will make angles in general with the corresponding lines in the diagram of simple liquid. The isodynamics of the diagram of the mixture, and those of the diagram of simple liquid, will differ in general in curvature at the line MM, but not in direction, $\frac{d\in}{dv} = -p$ for and $\frac{d\in}{d\eta} = t$.

The case is essentially the same with some substances, as water, for example, about the line which separates the simple liquid from a mixture of liquid and solid.

In these cases the inconvenience of having one diagram superposed upon another cannot be obviated by any change of the principle on which the diagram is based. For no distortion can bring the three sheets, which are united along the line MM (one on the left and two on the right), into a single plane surface without superposition. Such cases, therefore, are radically distinguished from those in which the superposition is caused by an unsuitable method of representation.

To find the character of a volume-entropy diagram of a perfect gas, we may make ε constant in equation (D), which will give for the equation of an isodynamic and isothermal

$\eta = a \log v + \text{Const.},$

And we may make p constant in equation (G), which will give for the equation of an isopiestic

$\eta = (a + c) \log v + \text{Const.}$

It will be observed that all the isodynamics and isothermals can be drawn by a single pattern and so also with the isopiestics.

The case will be nearly the same with vapours in a part of the diagram. In that part of the diagram which represents a mixture of liquid and vapour, the isothermals, which of course are identical with the isopiestics, are straight lines. For when a body is vapourized under constant pressure and temperature, the quantities of heat received are proportional to the increments of volume; therefore, the increments of entropy are proportional to the increments of volume.

As $\frac{d\epsilon}{dv} = -p$ and $\frac{d\epsilon}{d\eta} = t$, any isothermal is cut at the same angle by all the isodynamics, and is divided into equal segments by equidifferent isodynamics. The latter property is useful in drawing systems of equidifferent isodynamics.

The Isometric, Isopiestic, Isothermal and Isentropic

The arrangement of the isometric, the isopiestic, the isothermal and the isentropic drawn through any same point, in respect to the order in which they succeed one another around that point, and in respect to the sides of these lines toward which the volume, pressure, temperature and entropy increase, is not altered by any deformation of the surface on which the diagram is drawn, and is therefore independent of the method by which the diagram is formed. This arrangement is determined by certain of the most characteristic thermodynamic properties of the body in the state in question, and serves in turn to indicate these properties. It is determined, namely, by the value of $\left(\frac{dp}{d\eta}\right)_v$ as positive, negative, or zero, i.e., by the effect of heat as increasing or diminishing the pressure when the volume is maintained constant, and by the nature of the internal thermodynamic equilibrium of the body as stable or neutral,—an unstable equilibrium, except as a matter of speculation, is of course out of the question.

Let us first examine the case in which $\left(\frac{dp}{d\eta}\right)_v$ is positive and the equilibrium is stable. As $\left(\frac{dp}{d\eta}\right)_v$ does not vanish at the point in question, there is a definite isopiestic passing through that point, on one side of which the pressures are greater, and on the other less, than on the line itself. As $\left(\frac{dt}{dv}\right)_\eta = -\left(\frac{dp}{d\eta}\right)_v =$ -, the case is the same with the isothermal. It will be convenient to distinguish the sides of the isometric, isopiestic, etc., on which the volume, pressure, etc., increase, as the *positive* sides of these lines. The condition of stability requires that, when the pressure is constant, the temperature shall increase with the heat received,—therefore with the entropy. This may be written $[dt{:}d\eta]_p > 0$. It also requires that, when there is no transmission of heat, the pressure should increase as the volume diminishes, i.e., that $[dp{:}dv] < 0$. Through the point in question, A, let there be drawn the isometric vv' and the isentropic $\eta\eta'$, and let the positive sides of these lines be indicated as in the figure. The conditions $\left(\frac{dp}{d\eta}\right)_v > 0$ and $[dp{:}\,dv]_\eta < 0$ require that the pressure at v and at η shall be greater than at A, and hence, that the isopiestic shall fall as pp2 in the figure, and have its positive side turned as indicated. Again, the conditions $\left(\frac{dt}{dv}\right)_\eta < 0$ and $[dt{:}\,d\eta]_p > 0$ require that the temperature at η and at p shall be greater than at A, and hence, that the isothermal shall fall as tt2 and have its positive side turned as indicated. As it is not necessary that $\left(\frac{dt}{d\eta}\right)_p > 0$,, the lines pp2 and tt2 may be tangent to one another at A, provided that they cross one another, so as to have the same order about the point A as is represented in the figure; i.e., they may have a contact of the second (or any even) order. But the condition that $\left(\frac{dp}{d\eta}\right)_v > 0$, and hence $\left(\frac{dt}{dv}\right)_\eta < 0$, does not allow pp2 to be tangent to vv2, nor tt2 to $\eta\eta 2$.

If $\left(\frac{dp}{d\eta}\right)_v$ be still positive, but the equilibrium be neutral, it will be possible for the body to change its state without change either of temperature or of pressure; i.e., the isothermal and isopiestic will be identical. The lines will fall, except that the isothermal and isopiestic will be superposed.

In like manner, if $\left(\frac{dp}{d\eta}\right)_v < 0$, it may be proved that the lines will fall for stable equilibrium, and in the same way for neutral equilibrium, except that pp2 and tt2 will be superposed.

The case that $\left(\frac{dp}{d\eta}\right)_v = 0$ includes a considerable number of conceivable cases, which would require to be distinguished. It will be sufficient to mention those most likely to occur.

In a field of stable equilibrium it may occur that $\left(\frac{dp}{d\eta}\right)_v = 0$ along a line, on one side of which $\left(\frac{dp}{d\eta}\right)_v > 0$, and on the other side $\left(\frac{dp}{d\eta}\right)_v < 0$. At any point in such a line the isopiestics will be tangent to the isometrics and the isothermals to the isentropics. In a field of neutral equilibrium representing a mixture of two different states of the substance, where the isothermals and isopiestics are identical, a line may occur which has the threefold character of an isometric, an isothermal and an isopiestic. For such a line $\left(\frac{dp}{d\eta}\right)_v = 0$. If $\left(\frac{dp}{d\eta}\right)_v$ has opposite signs on opposite sides of this line, it, will be an isothermal of maximum or minimum temperature.

The case in which the body is partly solid, partly liquid and partly vapours has already been sufficiently discussed.

The arrangement of the isometric, isopiestic, etc., will indicate directly the sign of any differential co-efficient of the form $\left(\frac{du}{dw}\right)_z$, where u, w and z may be any of the quantities v, p, t, η (and ε, if the isodynamic be added in the figure). The value of such a differential co-efficient will be indicated, when the rates of increase of v, p, etc., are indicated, as by isometrics, etc., drawn both for the values of v, etc., at the point A, and for values differing from these by a small quantity. For example, the value of $\left(\frac{dp}{dv}\right)_\eta$ will be indicated by the ratio of the segments intercepted upon an isentropic by a pair of isometrics and a pair of isopiestics, of which the differences of volume and pressure have the same numerical value. The case in which W or H appears in the numerator or denominator instead of a function of the state of the body, can be reduced to the preceding by the substitution of pdv for dW, or that of $td\eta$ for dH.

In the foregoing discussion, the equations which express the fundamental principles of thermodynamics in an analytical form have been assumed, and the aim has only been to show how the same relations may be expressed geometrically. It would, however, be easy, starting from the first and second laws of thermodynamics as usually enunciated, to arrive at the same results without the aid of analytical formula,—to arrive, for example, at the conception of energy, of entropy, of absolute temperature, in the construction of the diagram without the analytical definitions of these quantities, and to obtain the various properties of the diagram without the analytical expression of the thermodynamic properties which they involve. Such a course would have been better fitted to show the independence and sufficiency of a graphical method, but perhaps less suitable for an examination of the comparative advantages or disadvantages of different graphical methods.

The possibility of treating the thermodynamics of fluids by such graphical methods as have been described evidently arises from the fact that the state of the body considered, like the position of a point in a plane, is capable of two and only two independent variations. It is, perhaps, worthy of notice, that when the diagram is only used to demonstrate or illustrate general theorems, it is not necessary, although it may be convenient, to assume any particular method of forming the diagram; it is enough to suppose the different states of the body to be represented continuously by points upon a sheet.

Chapter 14

Carnot Cycle

THE CARNOT CYCLE

All standard heat engines (steam, gasoline, diesel) work by supplying heat to a gas, the gas then expands in a cylinder and pushes a piston to do its work. The catch is that the heat and/or the gas must somehow then be dumped out of the cylinder to get ready for the next cycle. We examine the first step, the expansion, then go on to the full cycle—Carnot's analysis. Carnot's aim was to figure out how to maximize the efficiency of a heat engine, and then work out what that efficiency was, that is, how much of the heat supplied was actually converted into the mechanical work done by the engine. Remember that he had in mind the analogy of the water wheel, at that time still a main driving force of industry. He knew that the most efficient water wheels were those that operated smoothly, the water went into the buckets at the top from the same level, it didn't fall through any height, and didn't splash around. In the limit of a frictionless wheel, with gentle flow on and off the wheel, the machine would be reversible—turning it in reverse to raise the water back would take the same amount of work the wheel had delivered as the water fell. This was clearly perfect efficiency, so these were to conditions to emulate in the heat engine. The analog to having the water flow into buckets at the same height, with no wasteful drop, is to have the heat from the heat supply flow into the gas at the same temperature. There must of course be a slight drop in temperature for the heat to flow at all, but this must be minimized. This means that as the heat is supplied and the gas expands, the temperature of the gas stays the same as that of the heat supply (the "heat reservoir") and the gas is expanding *isothermally*.

ISOTHERMAL EXPANSION

So the first question is: how much work is done by an isothermally expanding gas? Taking the temperature of the heat reservoir to be T_H (*H* for hot), the expanding gas follows the isothermal path $PV = nRT_H$ in the (*P*, *V*) plane.

The work done by the gas in a small volume expansion ΔV is just $P\Delta V$, the area under the curve.Hence the work done in expanding isothermally from volume V_a to V_b is the total area under the curve between those values,

$$\text{work done isothermally} = \int_{V_a}^{V_b} P\,dV = \int_{V_a}^{V_b} \frac{nRT_H}{V}\,dV = nRT_H \ln\frac{V_b}{V_a}.$$

Since the gas is at constant temperature T_H, there is no change in its internal energy during this expansion, so the total heat supplied must be $nRT_H \ln\frac{V_b}{V_a}$, the same as the external work the gas has done.

In fact, this isothermal expansion is only the first step: the gas is at the temperature of the heat reservoir, hotter than its other surroundings, and will be able to continue expanding even if the heat supply is cut off. To ensure that this further expansion is also reversible, the gas must not be losing heat to the surroundings. That is, after the heat supply is cut off, there must be no further heat exchange with the surroundings, the expansion must be *adiabatic*.

ADIABATIC EXPANSION

The work done in an adiabatic expansion is like that done in allowing a compressed spring to expand against a force—equal to the work needed to compress the spring in the first place, for a perfect spring, and an adiabatically enclosed gas is essentially perfect in this respect. In other words, adiabatic expansion is reversible.

To find the work the gas does in expanding adiabatically from V_b to V_c, say, the above analysis is repeated with the isotherm $PV = nRT_H$ replaced by the adiabat $PV^\gamma = P_bV_b^\gamma$,

$$\text{work done adiabatically } W_{\text{adiabat}} = \int_{V_b}^{V_c} PdV = P_bV_b^\gamma \int_{V_b}^{V_c} \frac{dV}{V^\gamma} = P_bV_b^\gamma \frac{V_c^{1-\gamma} - V_b^{1-\gamma}}{1-\gamma}.$$

Again, this is the *area under the curve,* in this case under the adiabat, from b to c in the (P, V) plane.

Since points b, c are on the same adiabat, $P_cV_c^\gamma = P_bV_b^\gamma$, and the expression can be written more neatly:

$$W_{\text{adiabat}} = \frac{P_cV_c - P_bV_b}{1-\gamma}.$$

This is a useful expression for the work done since we are plotting in the (P, V) plane, but note that from the gas law $PV = nRT$, the numerator is just $nR(T_c - T_b)$, and from this $W_{\text{adiabat}} = nC_V(T_c - T_b)$, as of course it must be—this is the loss of internal energy that has been expended by the gas on expanding against external pressure.

We've looked in detail at the work a gas does in expanding as heat is supplied (isothermally) and when there is no heat exchange (adiabatically). These are the two initial steps in a heat engine, but it is equally necessary for the engine to get back to where it began, for the next cycle. The general idea is that the piston drives a wheel, which continues to turn and pushes the gas back to the original volume.

But it is also essential for the gas to be as cold as possible on this return leg, because the *wheel* is now having to expend work on the *gas,* and we want that to be as little work as possible—it's costing us. The colder the gas, the less pressure the wheel is pushing against.

To ensure that the engine is as efficient as possible, this return path to the starting point (P_a, V_a) must also be reversible. We can't just retrace the path taken in the first two legs, that would take all the work the engine did along those legs, and leave us with no net output. Now the gas cooled during the adiabatic expansion from *b* to *c*, from T_H to T_C, say, so we can go some distance back along the reversible colder isotherm T_C. But this won't get us back to (P_a, V_a), because that's on the T_H isotherm. The simplest option—the one chosen by Carnot—is to proceed back along the cold isotherm to the point where it intersects the adiabat through a, then follow that isotherm back to *a.* (One could follow a more complicated path: provided it was composed of segments each being adiabatic or isothermal, it would be reversible.) Carnot's cycle is around that curved quadrilateral having these four curves as its sides.

Efficiency of the Carnot Engine

In a complete cycle of Carnot's heat engine, the gas traces the path *abcd*. The important question is: what fraction of the heat supplied from the hot reservoir (along the red top isotherm) is turned into mechanical work? This fraction is called the *efficiency* of the engine.

The work output along any curve in the (*P*, *V*) plane is just $\int PdV$ —the area under the curve, but it will be *negative* if the volume is decreasing! So the work done by the engine during the hot isothermal segment is the area *abfh*, then the adiabatic expansion adds the area *bcef*, but as the gas is compressed back, the wheel has to do work on the gas equal to the area *cdge* as heat is dumped into the cold reservoir, then *dahg* as the gas is recompressed to the starting point.

The bottom line is that the total work done by the gas is the area bounded by the four paths: the curved "parallelogram" in the picture above. We could compute this area by finding $\int PdV$ for each segment, but that is unnecessary—on completing the cycle, the gas is back to its initial temperature, so has the same internal energy. *Therefore, the work done by the engine must be just the difference between the heat supplied at T_H and that dumped at T_C.*

Now the heat supplies along the initial hot isothermal path *ab*, equal to the work done along that leg, is (from the paragraph above on isothermal expansion):

$$Q_H = nRT_H \ln \frac{V_b}{V_a}$$

and the heat dumped into the cold reservoir along *cd* is

$$Q_C = nRT_C \ln \frac{V_c}{V_d}.$$

The difference between these two is the net work output. This can be simplified using the adiabatic equations for the other two sides of the cycle:

$$T_H V_b^{\gamma-1} = T_C V_c^{\gamma-1}$$

$$T_H V_a^{\gamma-1} = T_C V_d^{\gamma-1}.$$

Dividing the first of these equations by the second,

$$\left(\frac{V_b}{V_a}\right) = \left(\frac{V_c}{V_d}\right)$$

and using that in the preceding equation for Q_C,

$$Q_C = nRT_C \ln \frac{V_a}{V_b} = \frac{T_C}{T_H} Q_H.$$

The work done can now be written simply:

$$W = Q_H - Q_C = \left(1 - \frac{T_C}{T_H}\right) Q_H.$$

Therefore the efficiency of the engine, defined as the fraction of the ingoing heat energy that is converted to available work, is

$$\text{efficiency} = \frac{W}{Q_H} = 1 - \frac{T_C}{T_H}.$$

These temperatures are of course in degrees Kelvin, so for example the efficiency of a Carnot engine having a hot reservoir of boiling water and a cold reservoir ice cold water will be $1-(273/373) = 0.27$, just over a quarter of the heat energy is transformed into useful work.

THE SECOND LAW OF THERMODYNAMICS

After all the effort to construct an efficient heat engine, making it reversible to eliminate "friction" losses, etc., it is perhaps somewhat disappointing to find this figure of 27per cent efficiency when operating between 0 and 100 degrees Celsius. Surely we can do better than that? After all, the heat energy of hot water is the kinetic energy of the moving molecules, can't we find some device to channel all that energy into useful work? Well, we can do better than 27per cent, by having a colder cold reservoir, or a hotter hot one. But there's a limit: we can never reach 100per cent efficiency, because we cannot have a cold reservoir at $T_C = 0K$, and even if we did after the first cycle the heat dumped into it would warm it up!

The Second Law of Thermodynamics states that *we cannot devise an engine, working in a cycle, that simply extracts heat from a hot reservoir and delivers mechanical work.*

This means any engine that takes heat and delivers work also dumps out some of the initial heat to a reservoir at a lower temperature.

It's important to note that the First Law of Thermodynamics, the conservation of total energy including heat, would not be violated by an engine that powered a ship by extracting heat energy from the surrounding water. This Second Law is saying something new. And, this Second Law does not follow from the First by logical deduction—it comes (like the First) from experiment and observation.

THE MOST EFFICIENT ENGINE

An important consequence of the second Law is that *no engine can be more efficient than the Carnot cycle.* Essentially, this is because a "super efficient" engine, if one existed, could be used to drive a Carnot cycle in reverse, which would pump back to the hot reservoir the heat the super efficient engine dumped in the cold reservoir, and the net effect of the two coupled engines would be to take heat from the hot reservoir and do work, contradicting the Second Law.

To see this, we plot the heat/energy flow for the Carnot cycle:

Here $Q_H = Q_C + W$ (all expressed in Joules, of course).

Since the engine is reversible, it can also be run backwards (this would be a refrigerator: outside work is supplied, and heat is extracted from a cold reservoir and dumped into a hot reservoir:

Suppose now we have a super efficient engine, represented by the first diagram above, and dumping the same heat per cycle Q_C into the cold reservoir, but taking in more heat energy $Q_H + \Delta$ Joules from the hot reservoir, and performing work $W + \Delta$. Now, we hook up our super efficient engine to the "Carnot refrigerator" in the diagram above. The refrigerator sucks out of the cold reservoir all the heat the super efficient engine dumped there, and needs W Joules of work per cycle to do it. The super efficient engine can provide this, and there are still Δ Joules of work to spare. Of course, the Carnot refrigerator has also dumped Q_H Joules of heat in the hot reservoir. But the bottom line is that between them, the super efficient engine and the Carnot refrigerator have extracted Δ Joules of heat from the hot reservoir and performed Δ Joules of work—contradicting the Second Law.

The Second Law therefore forces the conclusion that no amount of machine design will produce an engine more efficient than the Carnot cycle. The rather low ultimate efficiencies this dictated came as a shock to nineteenth century engineers.

MOLECULAR COLLISIONS

In analyzing the gas so far, we've ignored collisions between molecules, and in fact for air at ordinary temperatures the relationship between pressure, volume and temperature came out correctly. Furthermore, Maxwell's speed distribution can be used to find what fraction of the molecules in a planet's atmosphere are moving at above escape velocity, so we can predict what gases will remain surrounding a planet, given the gravitational force near the surface, and the temperature. But there are other phenomena for which an understanding of collisions is all-important. For example, if two different gases, say oxygen and nitrogen, in a container at the same pressure and temperature are separated by a partition, how quickly will they mix once the partition is removed? Assuming room temperature, the molecules will be moving at hundreds of meters per second, so one might imagine the mixing will be over in hundredths of a second. But *that is not the case at all*—observationally, it might take an hour, for a box holding a few liters. This surprisingly slow penetration of one gas by another is called *diffusion*. The reason it takes the gases so long to mix becomes evident on tracking one molecule as it enters the other gas. Think of an oxygen molecule moving into nitrogen. We'll take O_2 and N_2 to be little spheres of diameter *d*. So now visualize the little O_2 sphere shooting into this space where all these other spheres are moving around. *Temporarily, for ease of visualization, let's imagine all the other spheres to be at rest.*

How far can we expect the O_2 to get before it hits an N_2? The *average* distance before a collision is called the *mean free path*. Let's try to picture how much room there is to fly between these fixed N_2 spheres. We do know that if it were *liquid* nitrogen, there would be *very little* room: liquids are just about incompressible, so the molecules must be touching. Roughly speaking, a molecule of diameter d will occupy a cubical volume of about d^3 (there has to be some space left over—we can pack cubes to fill space, but not spheres.) We also know that liquid nitrogen weighs about 800 kg per cubic meter, whereas N_2 gas at room temperature (and pressure) weighs about 1.2 kg per cubic meter, a ratio of 670. This means that *on average each molecule in the gas has 670 times more room*—that is, it has a space 670 times the volume d^3 we gave it in the liquid. So in the gas, the average center-to-center separation of the molecules will be the cube root of 670, which is about $8.75d$. So the picture is a gas of spheres of diameter d, placed at random, but separated on average by distances of order $10d$. It's clear that shooting an oxygen molecule into this it will get quite a way.

We now try to estimate just how far an O_2 will get, on average, as it shoots into this forest of spheres. Picture the motion of the center of the oxygen molecule. Before any collision, it will be moving on a straight-line path. Just how close does the O_2 center have to get to an N_2 center for a hit? Taking both O_2, N_2 to be spheres of diameter d, if an N_2 center lies within d of the O_2 center's path, there will be a hit. So we can think of the O_2 as sweeping out a volume, a cylinder of *radius d* centered on its path, hitting and deflecting if it encounters an N_2 *centered within that cylinder*. So how far will it get, on average, before a hit? In traveling a distance x, it sweeps out a volume $\pi d^2 x$. Now picture it going through the gas for some considerable length of time, so there are many collisions. The volume swept out will look like a stovepipe, long straight cylindrical sections connected by elbows at the collisions. The total volume of this stovepipe (ignoring tiny corrections from the elbows) will be just $\pi d^2 L$, L being the total length, that is, the total distance the molecule traveled. If the density of the nitrogen is *n* molecules per cubic meter, the number of N_2's in this stovepipe volume will be $\pi d^2 Ln$, in other words, this will be the number of collisions. Therefore, the average distance *between* collisions, the mean free path *l*, is given by:

$$\text{mean free path } l = \frac{\text{total distance traveled}}{\text{number of collisions}} = \frac{L}{\pi d^2 Ln} = \frac{1}{\pi d^2 n}$$

So what is *n*? We estimated above that each molecule has space $670d^3$ to itself, so *n* is just how many of those volumes there are in one cubic meter, that is, $n = 1/670d^3$.

Therefore, the mean free path is given by

$$l = \frac{1}{\pi d^2 n} = \frac{670d^3}{\pi d^2} = 200d.$$

Notice that this derivation of the mean free path *in terms of the molecular diameter* depends *only* on knowing the *ratio of the gas density to the liquid density—it does not depend on the actual size of the molecules*! But it *does* mean that if we can somehow measure the mean free path, by measuring how fast one gas diffuses into another, for example, we can deduce the size of the molecules, and historically this was one of the first ways the size of molecules was determined, and so Avogadro's number was found.

Let us now put in some numbers to find this mean free path: for O_2, N_2, $d = 0.3\text{nm}$, so the mean free path $l = 60\,\text{nm}$, or $6\times10^{-8}\text{m}$. The speed of the molecules at room temperature *v* is approximately 500 meters per sec., so the molecule has of order 10^{10} collisions per second!

Actually, there is one further correction we should make. We took the N_2 molecules to be at rest, whereas in fact they're moving as fast as the oxygen molecule, approximately. This means that even if the O_2 is temporarily at rest, it can undergo a collision as an N_2 comes towards it. Clearly, what really counts in the collision rate is the *relative* velocity of the molecules. Defining the average velocity as the root mean square velocity, if the O_2 has velocity $\vec{v}_1$ and the N_2 $\vec{v}_2$, then the square of the relative velocity $\overline{(\vec{v}_1 - \vec{v}_2)^2} = \overline{\vec{v}_1^2} - 2\overline{\vec{v}_1 \cdot \vec{v}_2} \cdot + \overline{\vec{v}_2^2} = \overline{\vec{v}_1^2} + \overline{\vec{v}_2^2}$, since $\overline{\vec{v}_1 \cdot \vec{v}_2}$ must average to zero, the relative directions being random. So the average *square* of the relative velocity is twice the average square of the velocity, and therefore the average root-mean-square velocity is up by a factor "2, and the collision rate is increased by this factor. Consequently, the mean free path is *decreased* by a factor of "2 when we take into account that *all* the molecules are moving.

Our final result, then, is that the mean free path

$$l = \frac{1}{\sqrt{2}\pi d^2 n}.$$

Finding the mean free path is—literally—the first step in figuring out how rapidly the oxygen atoms will diffuse into the nitrogen gas, and of course vice versa.

What we really want to know is just how much we can expect the gases to have intermingled after a given period of time. We'll just follow the one molecule, and estimate how far it gets. To begin with, let's assume for simplicity that it tales steps all of the same length *l*, but after each collision it bounces off in a random direction. So after *N* steps, it will have moved to a point

$$\vec{L} = \vec{l}_1 + \vec{l}_2 + \vec{l}_3 + \ldots + \vec{l}_N,$$

where each vector $\vec{l}_j$ has length *l*, but the vectors all point in random different directions.

If we now imagine many of the oxygen molecules following random paths like this, how far on average can we expect them to have drifted after *N* steps? (note that they could with equal likelihood be going backwards!) The appropriate measure is the root-mean-square distance,

$$\overline{\vec{L}^2} = \overline{\left(\vec{l}_1 + \vec{l}_2 + \vec{l}_3 + \ldots \vec{l}_N\right)^2} = Nl^2 + \sum_{i,j} \overline{\vec{l}_i \cdot \vec{l}_j}$$

Since the direction after each collision is completely random, $\overline{\vec{l}_i \cdot \vec{l}_j} = 0$, and the root-mean-square distance

$\sqrt{\overline{\vec{L}^2}} = \sqrt{N}l$.

If we allow steps of different lengths, the same argument works, but now l is the root-mean-square path length.

The important factor here is the $\sqrt{N}$. Recall from above that $l = 60\,\text{nm}$, or 6×10^{-8} m., and there are of order 10^{10} collisions per second. This means that the average distance diffused in one second is $\sqrt{10^{10}}\,l = 10^5 l$, say half a centimeter. The average distance in one hour would be only 60 times this, or 30 cm., one foot, and in a day about five feet—the average distance traveled is only increasing as the *square root* of the time elapsed!

This is a very general result. For example, suppose we have a gas in which the mean free path is l and the average speed of the molecules is v. Then the average time between collisions $\tau = l/v$. The number of collisions in time t will be t/τ, so the average distance a molecule moves in time t will be $r = l\sqrt{t/\tau}$.

A famous mystery cleared up by arguments like this was that Newton predicted the speed of sound would be given by $c^2 = B/\rho$, as we discussed earlier in the course, with B the bulk modulus. But when B was measured carefully by slowly compressing air, the result was in error by about 30per cent! The speed of sound predicted a higher (stiffer) bulk modulus.

The explanation turned out to be that in a slow measurement of the bulk modulus, the gas stays at the same temperature—the heating caused by slow compression leaks away. But if the compression is rapid, the gas heats up and so the pressure goes up more than if it had stayed at the same temperature. So the question is whether the compression and decompression as a sound wave passes through is so rapid that the heated-up gas doesn't have time to spread to the cooled regions. For sound at say 1000Hz, the wavelength is 34 cm. If compression heats gas locally, the hot molecules will diffuse away in a similar manner to that discussed above. They will be slightly faster than the average molecules. In 1/1000 th of a second, they will have 10^7 collisions, so will travel about $\sqrt{10^7}\, l = 3000l = 0.2\,\text{mm}$. This tiny distance compared with the wavelength of the sound wave means that during the compression/decompression cycles as the wave passes through, the heat has no chance to dissipate—so, effectively, it's like compressing a gas in an insulated container, it's harder to compress than it would be if the heat generated could flow away, and the bulk modulus is higher by an amount (around 30per cent) we shall work out in a forthcoming lecture.